Dictionary of Microscopy

Dictionary of Microscopy

Julian P. Heath

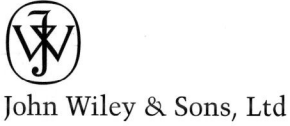

John Wiley & Sons, Ltd

Library of Congress Cataloging-in-Publication Data:
Heath, Julian P.
Dictionary of Microscopy / Julian P. Heath.
p. cm.
Includes bibliographical references and index.
ISBN-13: 978-0-470-01199-7 (cloth: alk. paper)
ISBN-10: 0-470-01199-8 (cloth: alk. paper)
1. Microscopy--Dictionaries. I. Title.
QH203.H43 2005
502'.8'203--dc22
2005016322

British Library Cataloguing in Publication Data:
A catalogue record for this book is available from the British Library
ISBN-13: 978-0-470-01199-7
ISBN-10: 0-470-01199-8

Typeset by Julian Heath in Times NewRomanPS and Frutiger.
Printed and bound in Great Britain by Micropress, Halesworth, Suffolk, UK.

Dedicated to
Geraldine, Caroline and Nicholas

Contents

Preface   9

Sponsors   11

Acknowledgements   13

Acronyms   15

Dictionary Entries   23

Bibliography   349

Sponsor Information   355

Preface

The past decade has seen huge advances in the development of microscopes and a corresponding broadening of their applications in the life and materials sciences. The modern microscope has now become a powerful analytical tool. Indeed, microscopes now have the capabilities equal to any biological, chemical or physical method of analysis but with the unique advantage of providing spatial information.

This welcome development in microscopy has been paralleled by an expansion of the vocabulary of technical terms used in microscopy: terms have been coined for new instruments and techniques and as microscopes reach even higher resolution the use of terms that relate to the optical and physical principles underpinning microscopy is now commonplace. The *Dictionary of Microscopy* was compiled to meet this challenge. The dictionary provides concise definitions of over 2,500 terms used in the fields of light microscopy, electron microscopy, scanning probe microscopy and X-ray microscopy. I hope this dictionary will be accessible to all microscopists, novice and experienced, and to scientists and laymen who need help to understand the microscopy lexicon.

The terms have been sourced from the microscopy literature, manufacturer's brochures and application notes, and from the web. In writing the definitions I have used as many published and web sources as I could find; the key sources are listed in the bibliography but I have also received generous assistance and written contributions from friends and colleagues whose names are listed in the acknowledgements. This dictionary is not exhaustive; the hardest part of writing a dictionary is knowing what to leave out and some readers will wonder why their favorite technique is absent. It is not possible to list every configuration or application of a particular microscope; on the other hand, for those microscopies that are described I have tried to put myself in the position of a complete novice and have included terms both simple and complex to help the understanding. Entries that seem blindingly obvious to an experienced microscopist can be useful to a novice or layman. Furthermore this is not an encyclopedia: the descriptions should be taken as an incentive and guide for further reading of the literature.

The entries are hierarchical: all of the terms used in the longer definitions are themselves defined elsewhere. Furthermore, the definitions are descriptive, not prescriptive: where there are synonymous terms - and there are many instances - I have listed the common ones and provided a cross-reference to the term that carries the definition; but this should not be taken as a recommendation for usage. For light microscopy terms, however, I have followed the guidelines set out by the Royal Microscopical Society Nomenclature Committee. In many other cases I did a web search to determine which synonym should carry the definition.

Microscopy is about images, so this Dictionary would be incomplete without the illustrations and technical diagrams that add context and information to the definitions. I have been very fortunate to have access to the *Microscopy and Analysis* article database and to the resources of the Sponsors to provide many of these illustrations.

Responsibility for the content and its accuracy is solely mine. I welcome any comments, corrections or suggestions for additional entries for future editions.

Julian P. Heath, Cambridge, July 2005

Sponsors

Gold Sponsors:

Gatan, Inc.

Leica Microsystems (UK) Ltd

Soft Imaging System

Thermo Electron Corporation

Carl Zeiss Ltd

Carl Zeiss SMT - Nano Technology Systems Division

Sponsors:

BAL-TEC AG

Boeckeler Instruments, Inc.

FEI Company

E.A. Fischione Instruments, Inc.

Hamamatsu Photonics

Leica Microsystems

Lindgren RF Enclosures, Inc.

Nikon Instruments

NT-MDT

Olympus Life and Material Science Europa GmbH

Oxford Instruments

PCO AG

Prior Scientific Instruments Ltd

Renishaw plc

StockerYale, Inc.

WITec GmbH

Cover illustrations:

Top right: energy-filtering transmission electron microscope (Carl Zeiss SMT); cryoTEM image (Carl Zeiss SMT); electron diffraction pattern (Carl Zeiss SMT).

Bottom right: cryopreparation and transfer system (Gatan); cryoSEM image (Gatan); CBED image (Gatan).

Bottom left: light microscope (Carl Zeiss); confocal fluorescence image (Carl Zeiss); phase contrast image (Julian Heath).

Top left: infrared microscope (Thermo); infrared image (Peter Lasch); Raman spectrum (WITec).

Cover design by Julian Heath.

Acknowledgements

This Dictionary could not have been written without the encouragement, advice and assistance of many friends and colleagues.

First, I would like to thank Jean and Peter Gordon of Rolston Gordon Communications, the former publisher of *Microscopy and Analysis,* for their enthusiastic support of this project from its inception; without their backing and hard work on the sponsorship and development stages, I would not have been able to fulfil my ambition of writing this dictionary.

The definitions were completed with the expert advice and contributions of many microscopists. I am extremely grateful to Peter Hawkes, who not only reviewed and commented on the first draft in its entirety but also contributed entries on light and electron optics. I am also grateful to David Williams, who supplied entries on electron optics and microanalysis, to Peter Evenett, who advised me on the writing style and the finer points of light microscopy, and to John Spence who contributed to the HREM entries. I thank Patrick Echlin, Alice Warley and Ian Watt for their reviews of and amendments to the first draft, Jack Vermeulen (Carl Zeiss SMT) for advice on electron microscopes, and Gerhard Holst (PCO AG) for advice on digital cameras. I also thank the many scientists who have collaborated with me or shared their expertise and wisdom with me during my time as a light and electron microscopist in Cambridge, London, Philadelphia and Houston, especially Audrey Glauert, Graham Dunn, Lee Peachey and Laszlo Komuves.

For the light microscopy and optics entries, Doug Murphy kindly allowed me to use many of the illustrations in his magisterial book *Fundamentals of Light Microscopy and Electronic Imaging* (John Wiley and Sons, Inc.), the second edition of which will be published in 2006 and is required reading for all microscopists.

I would also thank the sponsors of this dictionary and those contributors to *Microscopy and Analysis* whose illustrations I have used to add a richness to the content that I hope will help the reader better understand the text entries.

Finally, my thanks go to my colleagues at the publisher John Wiley & Sons, Ltd, especially my editor Jenny Cossham, and at the printer Micropress, for their invaluable support of this project and assistance with the editorial, layout and pre-press procedures.

Acronyms

This list contains many of the acronyms for microscopes and microscopical techniques that you may encounter in the microscopy literature. Please use this list as a guide to help you find an entry in the Dictionary, but note that not all of the acronyms are defined.

2D	two dimensional
2HM	second-harmonic microscopy
2PE	two-photon excitation
2PNIR	two-photon nearinfrared
3D	three dimensional
3DAP	three-dimensional atom probe
3HM	third harmonic microscopy
3PE	three-photon excitation
4D	four dimensional
4π	four pi
ADC	analog-to-digital converter
ADF	annular darkfield
ADF-STEM	annular darkfield scanning transmission electron microscope/microscopy
ADP	avalanche photodiode
AEM	analytical electron microscope/microscopy; Auger electron microscope/microscopy
AES	Auger electron spectroscopy
AFAM	atomic force acoustic microscopy
AFM	atomic force microscope/microscopy; adhesion force microscope/microscopy
AGC	automatic gain control
ALCHEMI	atom location by channeling-enhanced microanalysis
AMG	autometallography
AMLCD	active-matrix liquid crystal display
AOBS	acousto-optical beamsplitter
AOM	acousto-optical modulator
APFIM	atom probe field-ion microscope/microscopy
APM	atom probe microscope/microscopy
APS	active pixel sensor
ASA	American Standards Association
ASIC	application specific integrated circuit
ATR-IRM	attenuated total-reflectance infrared microscopy
ATW	atmospheric thin window
B&W	black and white
BAM	Brewster angle microscope/microscopy
BED	backscattered electron detector
BEEM	ballistic electron emission microscope/microscopy
BEI	backscattered electron imaging
BF	brightfield
BFP	back focal plane
BRET	bioluminescence resonance energy transfer

BSE	backscattered electron
BSED	backscattered electron detector/diffraction
CAFM	conductive atomic force microscope/microscopy
CALI	chromophore assisted laser inactivation
CARS	coherent anti-Stokes Raman spectroscopy
CB	coherent bremsstrahlung
CBED	convergent beam electron diffraction
CBLEED	convergent beam low-energy electron diffraction
CCD	charge-coupled device
CCM	charge-collection microscopy
CDF	centered darkfield
CFM	chemical force microscope/microscopy
CID	charge injection device
CITS	current imaging tunnelling spectroscopy
CL	cathodoluminescence
CLSM	confocal laser scanning microscope/microscopy
CMO	common main-objective
CMOS	complementary metal-oxide semiconductor
CMR	collosal magnetoresistance
CMYK	cyan magenta yellow key
CPD	critical point dryer/drying
CPM	coherence probe microscope/microscopy
CRT	cathode-ray tube
cryoEM	cryoelectron microscope/microscopy
cryoSEM	cryo scanning electron microscope/microscopy
cryoTEM	cryo transmission electron microscope/microscopy
CSLM	confocal scanning laser microscope/microscopy
CTEM	conventional transmission electron microscope/microscopy
CTF	contrast transfer function
CW	continuous wave
CWL	center wavelength
DF	darkfield
DFM	dynamic force microscope/microscopy
DFS	dynamic force spectroscopy
DFT	discrete Fourier transform
DIC	differential interference contrast
DIN	Deutsches Institut für Normung
DP	diffraction pattern
DPN	dip-pen nanolithography
DPOS	diffraction pattern observation screen
DQE	detector/detective quantum efficiency
DRIFTS	diffuse reflectance infrared Fourier-transform spectroscopy
DSTEM	dedicated scanning transmission electron microscope/microscopy
EBCCD	electron bombarded charge-coupled device
EBIC	electron beam-induced current/conductivity
EBSD	electron backscatter diffraction
EBSP	electron-backscatter pattern
ECCI	electron-channeling contrast imaging

ECP	electron-channeling pattern
ED	electron diffraction
EDX	energy-dispersive X-ray
EDXRF	energy-dispersive X-ray fluorescence
EDXS	energy-dispersive X-ray spectroscopy
EELS	electron energy-loss spectroscopy
EFI	energy-filtered imaging
EFL	effective focal length
EFM	electric force microscope/microscopy; electrostatic-force microscopy
EFTEM	energy-filtering transmission electron microscope/microscopy
ELFS	energy-loss fine structure
ELNES	energy-loss near-edge structure
EM	electron microscope/microscopy
EMCCD	electron-multiplying charge-coupled device
EMMA	electron-microscope microanalyzer
EPMA	electron-probe microanalyzer
EPXMA	electron-probe X-ray microanalysis
ESCA	electron spectroscopy for chemical analysis
ESEM	environmental scanning electron microscope/microscopy
ESI	electron spectroscopic imaging
ESPM	electrochemical scanning probe microscope/microscopy
ETEM	environmental transmission electron microscope/microscopy
EXAFS	extended X-ray absorption fine structure
EXELFS	extended energy-loss fine structure
FA	formaldehyde
FANSOM	fluorescence apertureless near-field scanning optical microscope/microscopy
FCS	fluorescent-cell sorting; fluorescence-correlation spectroscopy
FE	field emission
FEG	field-emission gun
FEGSEM	field-emission gun scanning electron microscope/microscopy
FEGTEM	field-emission gun transmission electron microscope/microscopy
FEM	field-emission microscope/microscopy
FET	field-effect transistor
FFM	friction force microscope/microscopy
FFT	fast Fourier transform
FIB	focused ion beam
FIB-SEM	focused ion-beam scanning electron microscope/microscopy
FIB-TEM	focused ion-beam and transmission electron microscopy
FIM	field-ion microscope/microscopy
FISH	fluorescent in-situ hybridization
FLAP	fluorescence localization after photobleaching
FLIC	fluorescence interference contrast microscopy
FLIM	fluorescence lifetime imaging microscopy
FLIP	fluorescence loss in photobleaching
FMM	force-modulation microscope/microscopy
FOLZ	first-order Laue zone
FP	fluorescent protein
FPR	fluorescence photobleaching recovery

FRAP	fluorescence recovery after photobleaching
FRET	fluorescence resonance energy transfer
FSM	fluorescent speckle microscopy
FT	Fourier transform
FTIRM	Fourier-transform infrared microscope
FWHM	full width at half maximum
FWTM	full width at tenth maximum
GA	glutaraldehyde
GE-EPMA	grazing-exit electron probe microanalysis
GIF	Gatan imaging filter
GIXPS	grazing incidence X-ray photoelectron spectroscopy
GSED	gaseous secondary electron detector
GSR	gunshot residue analysis
HAADF	high-angle annular darkfield
HAADFEM	high-angle annular darkfield electron microscopy
HFW	horizontal field width
HMC	Hoffmann modulation contrast
HOLZ	higher order Laue zone
HPGe	high-purity germanium
HREELS	high-resolution electron energy-loss spectroscopy
HREM	high-resolution electron microscope/microscopy
HRTEM	high-resolution transmission electron microscope/microscopy
HT	high tension
HV	high vacuum
HVEM	high-voltage electron microscope/microscopy
HWHM	half width at half maximum
I2M	image interference microscopy
I3M	incoherent-interference illumination
I5M	incoherent-interference illumination image interference
IETS	inelastic electron tunnelling spectroscopy
IFRAP	inverse fluorescence-recovery after photobleaching
IGSS	immunogold-silver staining
immunoEM	immunoelectron microscopy
IRM	interference reflection microscopy
ISH	in-situ hybridization
ISIT	intensifier silicon-intensifier target
IVEM	intermediate-voltage electron microscope/microscopy
IXRF	isotope X-ray fluorescence
K-M	Kossel-Möllenstedt
LACBED	large-angle convergent-beam electron diffraction
LASER	light amplification by stimulated emission of radiation
LCD	liquid-crystal display
LCTF	liquid-crystal tunable filter
LED	light-emitting diode
LEED	low-energy electron diffraction
LEEIXS	low-energy electron induced X-ray spectrometry
LEEM	low-energy electron microscope/microscopy
LEEPS	low-energy point source microscope/microscopy

LIMS	liquid-metal ion source
LM	light microscopy; low magnification
LSCM	laser scanning confocal microscope/microscopy
LSM	laser scanning microscope/microscopy
LVEM	low-voltage electron microscope/microscopy
LVSEM	low-voltage scanning electron microscope/microscopy
MAD	multiple-wavelength anomalous dispersion
MCA	multichannel analyzer
MDM	minimum detectable mass
MFM	magnetic force microscope/microscopy
MMF	minimum mass fraction
MOM	magneto-optical microscope/microscopy
MOS	metal-oxide semiconductor
MRFM	magnetic resonance force microscope/microscopy
MTF	modulation transfer function
NA	numerical aperture
NEXAFS	near-edge X-ray absorption fine structure microscopy
NIRRS	near-infrared reflectance spectroscopy
NSOM	nearfield scanning optical microscope/microscopy
NTSC	National Television Standards Committee
OAM	one-Angstrom microscope
OBIC	optical beam-induced current
OCT	optical coherence tomography
ODP	oil-diffusion pump
OD	optical density
OIM	orientation imaging microscopy
OPD	optical path(length) difference
OPL	optical pathlength
OTF	optical transfer function
PAL	phase-alternating line system
PCR	polymerase chain reaction
PEELS	parallel electron energy-loss spectrometer/spectroscopy
PEEM	photoemission electron microscope/microscopy
PEM	photoelectron microscope/microscopy
PET	positron emission tomography
PFM	photonic force microscope/microscopy
PFM	piezoforce microscope/microscopy
PFM	piezoresponse force microscope/microscopy
PIXE	proton-induced X-ray emission
PMT	photomultiplier tube
PPM	point-projection microscope/microscopy
PRI	piezoresponse imaging
PSD	position-sensitive detector
PSF	point-spread function
PSTM	photon scanning optical microscope/microscopy
QE	quantum efficiency
REBIC	remote electron beam-induced current
REELS	reflection electron energy-loss spectroscopy

REM	reflection electron microscope/microscopy
RF	radio/resonant frequency
RGB	red, green, blue
RHEED	reflection high-energy electron diffraction
RI	refractive index
RIMAPS	rotated image with maximum power spectrum
SACP	selected-area channelling pattern
SAD	selected-area diffraction
SAED	selected-area electron diffraction
SAM	scanning Auger microscope/microscopy
SATEM	sub-ångstrom transmission electron microscope
SCM	scanning capacitance microscope/microscopy
SDD	silicon drift detector
SE	secondary electron
SECAM	systéme electronique couleur avec mémoire
SECM	scanning electrochemical microscope/microscopy
SECPM	scanning electrochemical potential microscope/microscopy
SED	secondary electron detector
SEELS	serial electron energy-loss spectroscopy; surface electron energy-loss spectroscopy
SEI	secondary electron imaging
SEM	scanning electron microscope/microscopy
SEPM	scanning electrical potential microscope/microscopy
SERS	surface enhanced Raman microscope/microscopy
SESAM	sub-electronvolt sub-ångstrom microscope
SFB	surface force balance
SFM	scanning force microscope/microscopy
ShFM	shear force microscope/microscopy
SHPM	scanning Hall probe microscope/microscopy
SICM	scanning ion-conductance microscope/microscopy
SIM	scanning impedance microscope/microscopy
SIMS	secondary-ion mass spectrometry
SIT	silicon intensifier-target camera
SLM	spatial light modulator
S/N	signal-to-noise
SNOM	scanning near-field optical microscope/microscopy
SOLZ	second-order Laue zone
SPAM	scanning photoacoustic microscope/microscopy
SPELEEM	spectroscopic photoemission low-energy electron microscope/microscopy
SPLEEM	spin polarized low-energy electron microscope/microscopy
SPoM	surface-potential microscope/microscopy
SPPRM	surface plasmon polariton resonance microscope/microscopy
SQUID	superconducting quantum interference device
SRM	scanning resistance microscope/microscopy
SSEE	single sideband edge enhancement
SSRM	scanning spreading-resistance microscope/microscopy
STAP	scanning tunneling atom probe
STED	stimulated emission depletion
STEM	scanning transmission electron microscope/microscopy

SThM	scanning thermal microscope/microscopy
STM	scanning tunneling microscope/microscopy
STXM	scanning transmission X-ray microscope/microscopy
TDFM	transverse dynamic force microscope/microscopy
TEM	transmission electron microscope/microscopy
TEM-STEM	transmission electron microscope with scanning transmission capability
TFT	thin film transistor
THG	third harmonic generation
TIDF	tilted illumination darkfield
TIFF	tagged image file format
TIRF	total internal reflection fluorescence
TOA	take-off angle
TUNA	tunneling atomic force microscope/microscopy
UA	uranyl acetate
UHV	ultrahigh vacuum
UHVREM	ultrahigh-vacuum reflection electron microscopy
UHVSTM	ultrahigh-vacuum scanning tunnelling microscopy
UPS	ultraviolet photoelectron spectroscopy
UTW	ultrathin window
UV	ultraviolet
VA-TIRFM	variable-angle total internal reflection fluorescence microscopy
VCSEL	vertical cavity surface-emitting laser
VPSEM	variable-pressure scanning electron microscope/microscopy
WBDF	weak-beam darkfield
WDXS	wavelength-dispersive X-ray spectrometry/spectroscopy
WDX	wavelength-dispersive X-ray
WDXRF	wavelength dispersive X-ray fluorescence
XEDS	X-ray energy-dispersive spectrometry/spectroscopy
XPS	X-ray photoelectron spectroscopy
XRD	X-ray diffraction
XRF	X-ray fluorescence
XRM	X-ray microscope/microscopy
XSW	X-ray standing wave
XTEM	cross-section(al) transmission electron microscopy
YAG	yttrium aluminum garnet
YAP	yttrium aluminum perovskite
ZAF	atomic number (Z), absorption (A), fluorescence (F)
ZAP	zone-axis pattern
ZLP	zero-loss peak
ZOLZ	zero-order Laue zone

Abbe condenser

A high numerical aperture light microscope condenser, typically containing two uncorrected lenses suitable for most objectives.

Abbe constant see Abbe number

Abbe number

A measure of the dispersion of a transparent material. Symbol v or V. Abbe number is defined as: $v = (n_d - 1)/(n_F - n_C)$, where n is refractive index at wavelengths of Fraunhofer d, F and C lines.

Abbe sine condition see sine condition

Abbe theory of image formation

A theory for the formation of an image in a diffraction-limited light microscope formulated by Ernst Abbe. Abbe's theory states that image contrast is generated by interference of zero and higher order diffracted beams in the image plane. The degree of detail in the image is determined by the number of higher order diffracted beams captured by the objective lens.

aberration

The degradation of a perfect image arising from the materials, manufacture or operation of an optical system. The formation of a perfectly sharp, point-to-point image by a lens is a first-order or paraxial approximation.

Images are degraded by aberrations of three kinds: monochromatic, chromatic and parasitic aberrations. Mono-chromatic aberrations are caused by departures of lens properties from the paraxial approximation. Chromatic aberration arises from the dependence of the lens properties on the wavelength of the radiation. Parasitic aberrations are the consequence of mechanical or similar imperfections in lens construction, particularly in electron lenses.

Abbe condensers. Courtesy of Carl Zeiss Ltd.

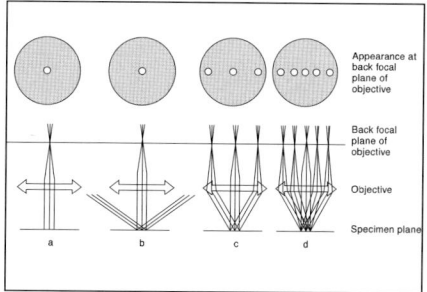

Abbe theory of image formation. (a, b) Zero order light capture only: no image is formed. (c) Zero plus first order capture: minimum condition for image formation. (d) Capture of higher orders leads to increased resolution. From D. Murphy © John Wiley & Sons, Inc.

aberration corrector

Magnetic or electrostatic lenses that correct aberrations in electron microscopes and electron spectrometers. Axial astigmatism and axial comas are corrected by stigmators or more complex multipole pole devices. Spherical aberration can be corrected with a quadrupole-octopole corrector for STEMs or a sextupole corrector for TEMs. Chromatic aberration correction requires either hybrid magnetic-electrostatic quadrupole lenses or a round lens-quadrupole unit.

absorbance see optical density

absorption

The trapping of incident radiation by a specimen so that there is no further propagation; the energy of the radiation is converted to heat or other forms of energy.

absorption band

A discontinuous change of intensity in a spectrum caused by the absorption of radiation by the transmitting or emitting medium.

absorption contrast see amplitude contrast

absorption correction

A correction factor used in quantitative X-ray microanalysis to account for the absorption of X-rays by the specimen. Absorption correction is used with thick specimens and with thinner specimens whenever X-rays travel at oblique angles through the specimen before reaching the detector.

absorption edge

A discontinuous change of intensity in an X-ray microanalysis emission spectrum caused by the absorption of X-rays by the specimen or by components of the detector system.

absorption maximum

The energy or wavelength of radiation that is maximally absorbed by a material.

accelerating voltage

The potential difference between the filament and the anode in an electron microscope. Symbols: V, kV, MV; units: volts. Accelerating voltage determines the wavelength and energy of an electron, and influences specimen penetration and image resolution. The relativistic accelerating voltage Φ_r is given by the expression:
$$\Phi_r = \Phi(1 + e\Phi)$$
where Φ is nominal voltage and $e = 0.978$ MV^{-1}.

acceleration voltage see accelerating voltage

accelerator

1. A chemical catalyst that increases the rate of polymerization. Accelerators are added to embedding resins: e.g. benzyldimethylamine or *tris*-dimethylaminomethyl phenol for Araldite and Epon; dimethylaminoethanol for Spurr's resin; and dibenzoyl peroxidase for LR White resin.
2. A device that accelerates charged particles.

acceptance angle

The solid angle subtended by the margins of the collimator aperture and the centre of the detector of an X-ray spectrometer.

accommodation

The ability to change the refractive power of the eye lens to focus on objects near or far. Accommodation is mediated by the ciliary muscles.

achromat

1. A light microscope objective with minimal chromatic aberration at two wavelengths, typically blue and red, and minimal spherical aberration for green light.
2. A lens with similar specification.

achromatic-aplanatic condenser

A high numerical aperture light microscope condenser, corrected for chromatic and spherical aberrations and coma, typically used with high magnification objectives.

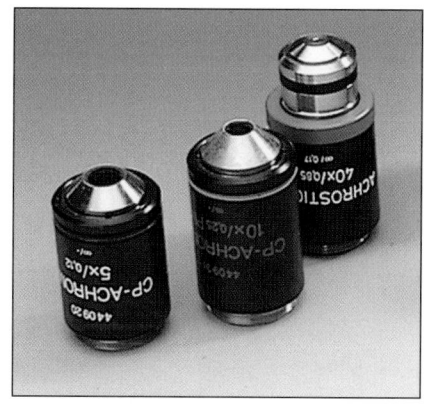

Achromat objectives. Courtesy of Carl Zeiss Ltd.

achromatic condenser

A medium numerical aperture light microscope condenser, corrected for chromatic aberration, and suitable for use with most dry objectives.

achromatic objective see achromat

acoustic microscope

A microscope that uses a focused beam of ultrasound waves, typically in the MHz range, to measure and map the transmission of ultrasound waves through a specimen. A standard configuration is the reflection-mode acoustic microscope in which a piezoelectric ultrasound transducer rapidly switches between send and receive modes, sending an ultrasound beam into the specimen and receiving its echoes. The transducer is coupled to the specimen by a fluid, usually water. The sound waves are sensitive to and reflected by changes in acoustic impedance of the specimen due to density and defects, such as cracks and voids. The return echoes can be gated to select those from a specific depth within the specimen.

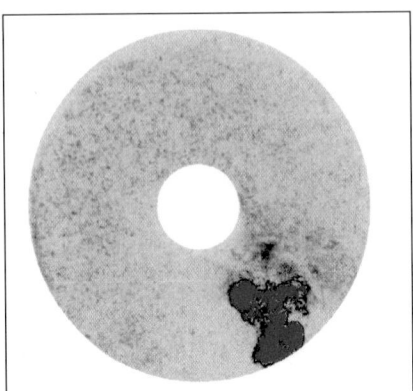

Localization of defects in ceramic by acoustic microscopy. *Microscopy and Analysis* 2003.

acoustic microscopy

The use of an acoustic microscope.

acousto-optical beam scanner see acousto-optic beamsplitter

acousto-optical beamsplitter

A beamsplitter whose optical properties are determined by acoustic regulation of the refractive index of a crystal. An acousto-optical beam splitter can be used as a wavelength selecting filter or as a beam scanning device. A typical acousto-optical beamsplitter is made of a crystal of tellurium dioxide attached to a piezo transducer; as the piezo vibrates it compresses the crystal, changing its RI and hence the degree of refraction of light passing through the crystal. This property can be used to select a wavelength of emission or excitation light, or to deflect or scan a beam of monochromatic light.

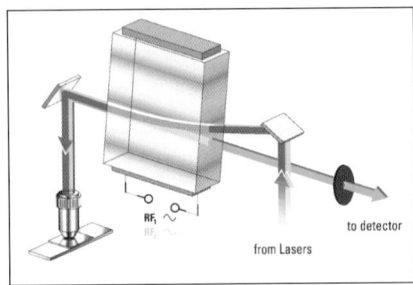

Acousto-optical beamsplitter. Courtesy of Leica Microsystems.

acousto-optical deflector/modulator see acousto-optic beamsplitter

acrylic resin

A synthetic resin prepared by polymerizing esters or other derivatives of acrylic acid (propenoic acid). Acrylic resins are low viscosity resins, partially water miscible, and can be polymerized by heat or UV light at high and low temperatures. Examples: Lowicryl, Unicryl, LR White, LR Gold. Acrylic resins are commonly used for cyto-chemical studies of sectioned specimens.

active layer

The intrinsic region of a silicon detector that generates measurable charges.

active-matrix liquid crystal display see thin-film transistor display

active-pixel sensor

A photodiode array in which each pixel has a transistor for signal amplification and processing. Active pixel sensors are commonly manufactured using CMOS technology.

adaptation see dark adaptation

adaptive optics

The correction of image aberrations by using a deformable mirror to reflect light from an optical system onto a wavefront sensor; the sensor analyzes the image for aberrations and uses its information to change the geometry of the mirror to minimize aberrations.

additive colors

Any set of colors that added together create white. The additive colors used in most imaging devices are red, green and blue.

adhesion force microscopy

A type of scanning probe microscopy that measures the forces of adhesion between the probe and the specimen. Electrical, magnetic or interatomic forces are measured by collecting a force curve, which is a plot of cantilever deflection as a function of probe-sample distance. A complete force curve includes forces measured while the tip is approaching and retracting from the surface. Typically there is hysteresis due to some sort of

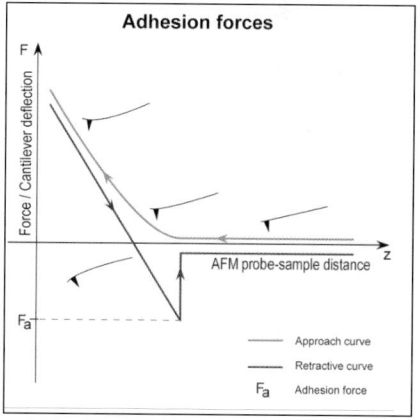

Principle of adhesion force microscopy. Courtesy of NT-MDT.

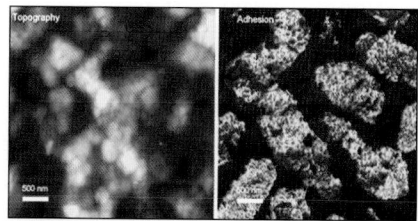

Adhesion force microscopy (topography left; adhesion force mode right) of a stone containing fossilized bacteria. The bacteria are only visible in the adhesion image. Courtesy of WITec.

adhesion which appears in the force curve as a deflection below the zero-deflection line. The most negative force detected during the retraction is regarded as the adhesion force.

adsorption
The binding of molecules and atoms to a surface.

airlock
An evacuatable chamber in a device maintained under vacuum that allows the introduction of material or exchange of components without degradation of the system vacuum; for example, the specimen holder and gun airlocks of an electron microscope.

Airy disk
The central spot in the Airy pattern of a bright point imaged by an optical system. The radius of the Airy disk is $0.61\lambda/NA$, where λ is wavelength and NA is numerical aperture.

Airy pattern
The image of a point object formed by a diffraction-limited microscope. The Airy pattern comprises an Airy disk surrounded by rings of decreasing intensity called the side lobes; the Airy disk contains ~84% of the energy; the first ring ~7%; the second ~3%; and the third ~1.5%.

Airy resolution see Rayleigh criterion

aliasing
The false periodicity of detail in an image of a specimen with regularly spaced elements produced by a digital imaging device that has periodic recording units, i.e. pixels. Aliasing occurs when the imaging device samples the object below the Nyquist criterion. Aliasing can usually be eliminated by increasing magnification so that at least two pixels in the imaging device sample each periodicity in the specimen.

amorphous ice see vitreous ice

amplitude contrast
Contrast in an image that is derived from the reduction in the amplitude of transmitted radiation

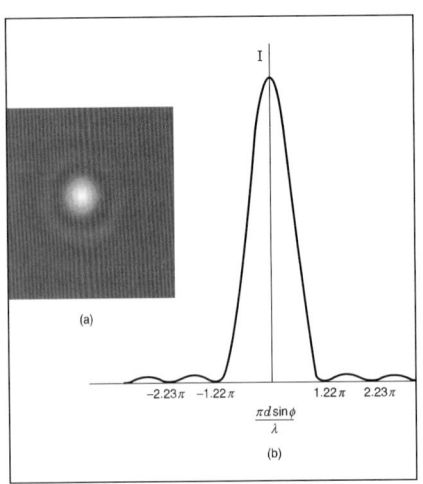

Airy pattern of point source (a) and plot of intensity (b) in a diffraction-limited optical system. From D. Murphy © John Wiley & Sons, Inc.

due to absorption, diffraction or scattering. Amplitude contrast is the major mechanism of image contrast in light and electron microscopy of thick, stained or amorphous specimens.

anaglyph
A composite image formed by combining differently colored versions of the two images of a stereo pair.

anaglyph glasses/spectacles
A pair of glasses (spectacles) containing differently colored filters for each eye, used for viewing anaglyphs.

analog-to-digital converter
An electronic device that converts a continuous signal to a series of discrete signals. Analog-to-digital converters are used to digitize signals from charge-coupled devices and spectrometers. In an X-ray spectrometer the ADC produces a series of constant height pulses proportional in number to the amplitude of each signal pulse received from the pulse processor.

analytical electron microscope
A transmission electron microscope capable of forming a small (<1 nm) electron probe and equipped with detectors for electron energy-loss and energy-dispersive spectroscopy for high spatial resolution chemical analysis of thin specimens.

analytical electron microscopy
The use of an analytical electron microscope.

analyzer
A polarizing filter used to determine the plane of vibration of polarized light. An analyzer is usually placed between the objective and intermediate image plane in DIC, interference and polarized light microscopes.

anastigmat
A lens corrected for astigmatism and field curvature.

anastigmatic
Having no astigmatism.

Red-green anaglyph image of salt crystals. Courtesy of Soft Imaging System.

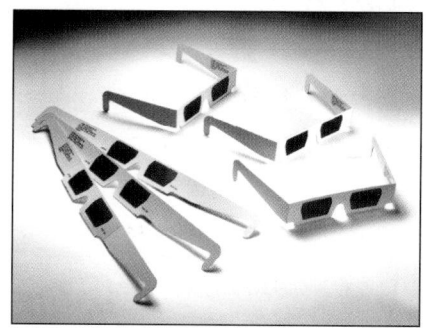

Anaglyph glasses. Courtesy of Agar Scientific.

anastigmatic objective see **anastigmat**

angle of incidence
The angle formed by a ray and a normal (perpendicular line) at the point of incidence.

ångström
A non-SI unit of length. Symbol: Å. One ångström is equal to 10^{-10} meter, or 0.1 nanometer. The ångström is widely used in electron and X-ray microscopy as it is similar to the average atomic diameter.

angular magnification
1. The ratio of the angles subtended by the image and object at the eye when viewed at the reference viewing distance.
2. In electron microscopy, for small angles, the ratio of the semi-angle subtended at the exit pupil by an axial point in the image to the semi-angle subtended at the entrance pupil by the same point in the object. Hence angular magnification is the reciprocal of lateral magnification.

angular resolution
The angle subtended by the minimum resolvable distance at a lens. Using Rayleigh's criterion, the minimum resolvable angular separation of a lens is equal to $1.22\lambda/D_{ep}$, where λ is wavelength and D_{ep} is the diameter of entrance pupil. For the human eye, the angular resolution is about 1 minute of arc, that is approximately 1 mm at 3 m.

annular darkfield detector
An annular solid-state detector for elastically scattered electrons, used in darkfield imaging mode in scanning transmission and transmission electron microscopes. The detector has a central aperture for the direct beam or a bright field detector and typically detects electrons scattered though semiangles of 10-50 mrad (0.5-3°).

annular darkfield mode
The use of an annular darkfield detector for the detection of elastically scattered electrons in a scanning transmission or transmission electron microscope.

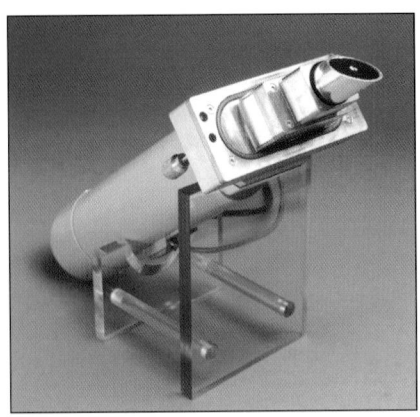

Annular darkfield detector. Courtesy of E.A. Fischione Instruments, Inc.

annular diaphragm

A diaphragm with a transparent annulus that transmits a hollow cone of light. Annular diaphragms are used in condensers for phase contrast and darkfield light microscopy.

annulus

A ring-shaped object or aperture.

anode

The more positively charged terminal of an electrical device.

anode plate

A metal plate with a central circular aperture that attracts electrons from an electron gun into the column of an electron microscope. The anode plate has a positive potential (the accelerating voltage) relative to the gun cathode (filament), attracting electrons into the first condenser lens.

anodic tip oxidation

A method of nanolithography using a scanning probe microscope in which a negatively biased metal tip is used to induce oxidation of a metal surface. Anodic tip oxidation requires a humid atmosphere to form a water droplet between tip and substrate. Hydroxyl ions produced by the current flowing through the droplet oxidize the substrate, building nanometer-sized metal oxide structures as the tip moves over the specimen.

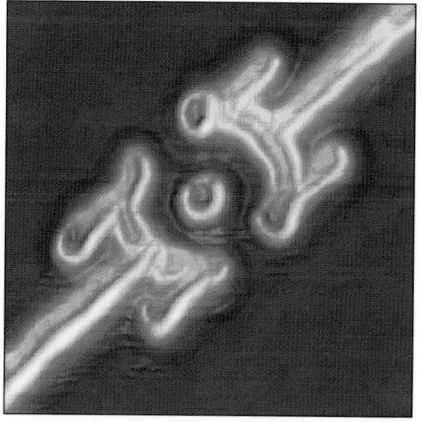

AFM image of quantum interferometer produced by anodic tip-induced oxidation of AlGaAs/ GaAs heterostructure, with inner radius ~60 nm and channel width ~15 nm. Image courtesy of NT-MDT and D. Sheglov, Z. Kvon, A.Toropov and A. Latyshev, Institute of Semiconductor Physics, Novosibirsk, Russia.

anomalous dispersion

The decrease in refractive index with increasing energy, due to absorption of higher energy radiation.

antibody

A protein produced by the immune system of an animal that binds to an epitope of another molecule (antigen). Antibodies are used as primary labels in immunocytochemistry.

anticontaminator

A device that reduces contamination of the specimen and internal surfaces of an electron microscope.

antifade medium

A mounting medium that reduces the photo-bleaching of fluorophores. Antifade media typically contain ingredients that neutralize free radicals, such as reactive oxygen species.

Antiflex objective

Proprietary name for a light microscope objective with a quarter wave plate in front of the front lens, typically used for reflection contrast microscopy.

antigen retrieval

The treatment of specimens to enhance the affinity of cytochemical probes for their receptors. Antigen-retrieval procedures typically include a combination of chemical, enzymatic and heat treatments to remove embedding matrix and cell components and to reverse the effects of chemical fixation and dehydration.

antireflection coating

A thin film of material deposited on the surface of a lens that reduces the reflection of light by destructive interference of light reflected from the two surfaces of the film. Anti-reflection coatings are typically made of magnesium fluoride, dielectrics or metals.

antistatic device

A device that produces a stream of ionized particles that neutralizes charged surfaces and instrument parts. Antistatic devices may be used to reduce the hydrophobicity of specimen supports and in cryoultramicrotomy to prevent sections moving away from the knife.

anti-Stokes Raman scattering

Raman scattering of molecules in an excited state leading to emission of photons of higher energy and shorter wavelength.

anti-Stokes shift

The increase in frequency (and decrease of wavelength) of emitted versus incident light that occurs during luminescence or scattering events. The anti-Stokes shift is a non-linear optical phenomenon.

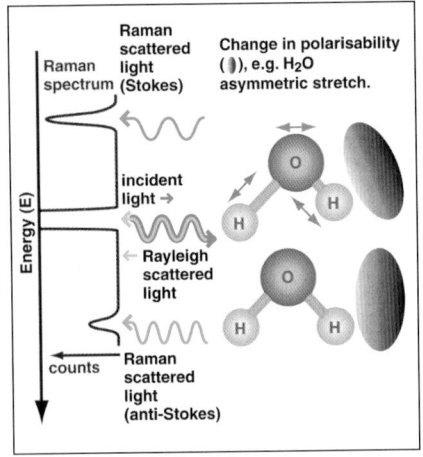

Principle of anti-Stokes Raman scattering. Courtesy of Renishaw plc.

antivibration table

A platform for instruments and microscopes that dampens physical vibrations from outside sources.

apertometer

An instrument for measuring the numerical aperture of a light microscope objective.

aperture

A hole or opening in an optical system or detector through which radiation or matter passes. In light microscopy the term aperture is commonly used to describe the opening of a lens. In electron microscopy the term aperture is commonly used to describe a diaphragm, such as the condenser aperture or pressure-limiting aperture.

aperture aberration see spherical aberration

aperture angle

The angle subtended by the margins of an aperture or lens at a point in the specimen or image.

aperture contrast

Contrast in an image that derives from the use of an aperture diaphragm to reduce the cone of rays or electrons that form the image. In a light microscope, closing the condenser iris diaphragm increases contrast. In a TEM, the objective aperture provides contrast by blocking high-angle scattered electrons.

apertureless near-field scanning optical microscopy

The use of a near-field scanning optical microscope with a probe that has no aperture; the specimen is illuminated by light reflected from the tip of the probe, which is placed at the focus of a laser beam.

aperture mode see conoscopic mode

aperture near-field scanning microscopy

The use of a near-field scanning optical microscope with a probe that has an aperture through which the specimen is illuminated.

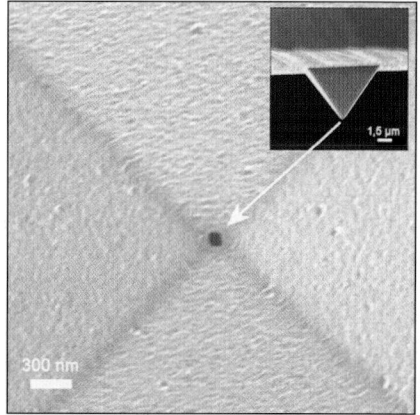

NSOM aperture at the tip of a hollow pyramid on a silicon cantilever. Courtesy of WITec.

Apochromat objectives. Courtesy of Nikon.

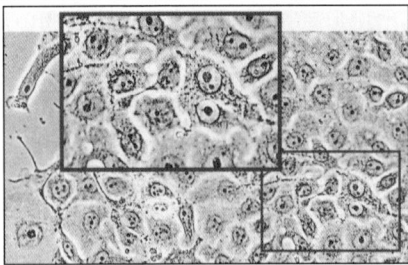

Comparison of images of cultured cells with standard (upper) and apodized (lower) phase contrast objectives. Courtesy of Nikon.

aperture planes see **conjugate aperture planes**

aperture scanning phase-contrast microscopy

A form of phase contrast light microscopy in which the condenser annulus is replaced by a circulating narrow beam of light that is conjugate with a synchronously rotating phase-shifting disk in the back focal plane of the objective.

aperture stop

Any object, such as a diaphragm or the mount of a lens, that regulates the amount of light that forms an image.

aplanatic

A definition of an optical system corrected for spherical aberration and coma by satisfying the Abbe sine condition.

aplanatic condenser

A high numerical aperture light microscope condenser, corrected for spherical aberration and coma, typically used with high magnification objectives.

apochromat

1. A light microscope objective corrected for chromatic aberration at up to four wavelengths (e.g. UV, blue, green, red) and for spherical aberration at two or more wavelengths.
2. A lens of similar specification.

apochromatic objective see apochromat

apodization

The reduction of the side lobes of the Airy disk, typically accomplished by the use of structured illumination.

apodized objective

A type of phase contrast objective that reduces the bright halos commonly seen at the edges of thick objects. An apodized objective contains a phase plate with neutral density rings either side of the phase annulus to reduce the intensity of any diffracted light that passes close to the annulus.

apodized phase-contrast objective see apodized objective

application-specific integrated circuit detector

A semiconductor detector with a photodiode array mounted on an integrated circuit which processes the signals from each pixel using electronic circuitry designed specifically for the application.

aqueous humor

The watery fluid filling the anterior chamber of the eye. Refractive index = 1.336.

Araldite

Proprietary name for commonly used epoxy resin embedding medium.

arc lamp

A gas-discharge lamp with two pointed tungsten electrodes in a quartz glass bulb. An electric arc is produced when a voltage of several thousand volts is applied to the electrodes. The intense heat of the arc creates a high-pressure plasma from elements inside the bulb, such as mercury or xenon, which generate the characteristic radiation.

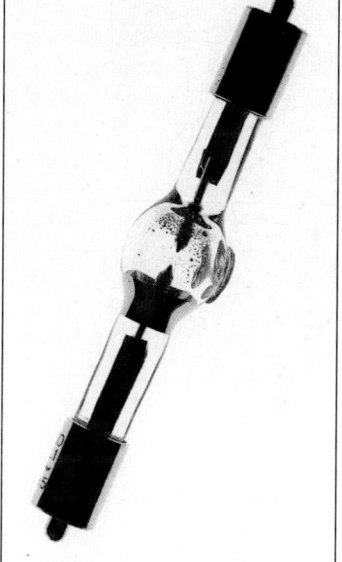

Arc lamp. Courtesy of Carl Zeiss Ltd.

arm see microscope arm

ASA film rating

The American Standards Association standard film speed, now superseded by the similar ISO rating.

aspect ratio

The ratio of height to width.

aspheric lens

A lens with one or both surfaces that are not truly part of a sphere, typically for the correction of spherical aberration.

astigmatism

An aberration of lenses that causes rays in a plane parallel to the optical axis to be focused at a different focal point from rays in an orthogonal plane. Astigmatism causes a circular object to be imaged as an ellipse. Astigmatism depends linearly on angle and quadratically on off-axis distance at

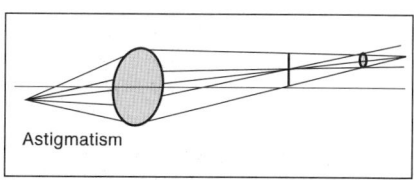

Astigmatism. From D. Murphy © John Wiley & Sons, Inc.

the object plane.

asymmetric objective lens

In electron optics, any lens in which the magnetic field or potential distribution on the optical axis is not symmetric about a mid-plane

asymptotic cardinal elements

In electron optics, the cardinal elements that characterize condensers, intermediate and projector lenses, for which there is no real object or image inside the field.

atmosphere

A unit of pressure. Symbol atm. One atmosphere equals 101,325 pascals.

atmospheric thin window

A window in an X-ray spectrometer designed to withstand atmospheric pressure. The window is usually made of a thin film of boron or silicon nitride, diamond or polymer.

atom lens

A lens formed by a column of heavy atoms. An atom lens has extremely small aberrations and can be used to refocus a beam of electrons in high resolution electron microscopy.

atom location by channeling-enhanced microanalysis

A technique used to locate substitutional impurity atoms in thin crystals using an energy-dispersive spectrometer in a transmission electron microscope. In ALCHEMI the characteristic X-ray emission from the impurity is compared with that from host atoms as the diffraction conditions are varied.

atom mirror

A silicon crystal that focuses a beam of atoms by changing the electric field in the crystal.

atom probe field-ion microscope see three-dimensional atom probe microscope
atom probe microscopy see three-dimensional atom probe microscopy

atomic-column EELS

The acquisition of electron energy-loss spectra from a crystalline specimen, oriented such that a small (< 0.5 nm) STEM probe is channeled along an individual column of atoms. Thus all spectroscopic information is obtained at atomic resolution and is directly comparable with the atomic-resolution Z-contrast image obtained at the same time.

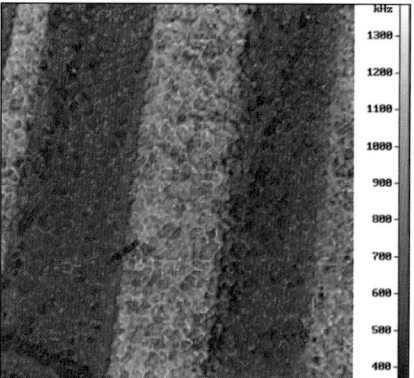

atomic force acoustic microscopy

The use of an atomic force microscope to measure and map acoustic waves transmitted through a specimen. In AFAM the specimen is placed on a piezoelectric acoustic wave generator; vibrations at the sample surface are investigated by placing a cantilever tip in contact with the specimen. AFAM provides contrast imaging of specimen hardness and can determine Young's modulus at each point of the scan.

Young's modulus map of low and high density polyethylene with different elasticity, obtained by atomic force acoustic microscopy. Scan size: 47 x 47 μm. Courtesy of NT-MDT.

atomic force microscope

A collective term for a family of scanning probe microscopes that measure and map the attractive and repulsive forces between a sharp probe and the specimen. The term derives from the ability of the AFM to measure forces on the scale of those between atoms.

atomic force microscopy

The use of an atomic force microscope.

atomic number

The number of protons in the nucleus of an atom, or the number of electrons in a neutral atom. Symbol Z.

atomic number contrast see Z contrast
atomic scattering amplitude see atomic scattering factor

atomic scattering factor

A measure of the amplitude of an electron wave scattered by an atom. Symbol $f(\theta)$. The atomic scattering amplitude factor is proportional to the atomic number of the scattering atom and is useful for describing low-angle elastic scattering which decreases rapidly with increasing angle of

Atomic force microscope. Courtesy of WITec.

scattering away from the optic axis of the microscope. The intensity of scattering is proportional to the square of $f(\theta)$.

attenuated total-reflectance infrared microscopy

A type of infrared microscopy that analyses the specimen by total internal reflection of the probing beam through a crystal in contact with the specimen. In ATR-IM light is directed along the crystal at an angle that causes multiple internal reflections from the crystal-specimen interface, producing evanescent waves that enter and are absorbed by the specimen, reducing the intensity of the incident IR beam. On leaving the crystal the transmitted beam is then analyzed with a detector or spectrometer to produce a map or spectrum of infrared absorption.

attenuated total-reflectance infrared objective

A light microscope objective for attenuated total-reflectance infrared microscopy with a crystal (typically diamond) mounted in front of the lens and optics that direct an infrared beam through the objective into the crystal and back to the detector. The crystal is placed in direct contact with the specimen.

Auger electron

An outer shell electron released during the excitation of an atom. An atom releases an Auger electron when incident radiation ejects an inner-shell electron, creating a hole that is filled by an electron from an intermediate shell; this process releases energy that is used to eject an Auger electron from a higher shell.

Auger electron spectroscopy

Compositional analysis of the surface layers of a specimen using Auger electrons released by an incident electron beam. In Auger electron spectroscopy the specimen is probed with a focused electron beam (1-20 kV) and Auger electrons are detected and analyzed by an electron spectrometer. AES is always performed in an ultrahigh vacuum to reduce surface contamination; sputtering with an inert gas may be used to remove

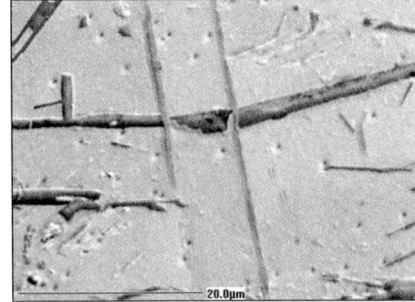
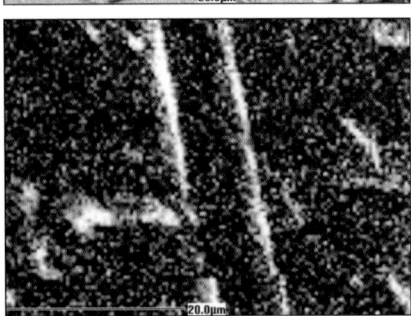

Auger electron spectroscopy. (a) SEM of nickel superalloy. (b) AES titanium map. *Microscopy and Analysis* 2000.

surface contamination and for depth profiling. AES is suitable for low Z elements that cannot be easily detected by X-ray microanalysis.

autocorrelation

The cross-correlation of an image or signal with itself. The original function is displaced by a small amount in time or space and the two signals compared by cross-correlation. The autocorrelation function is defined by:

$$R(x^1, y^1) = \iint f(x, y) f^* (x - x^1, y - y^1) \, dx \, dy$$

where * denotes the complex conjugate. Owing to the close relation to a convolution, the Fourier transform of R is equal to $|F|^2$ where F is the Fourier transform of f.

autofluorescence

The fluorescence of a specimen arising from endogenous fluorophores or specimen processing.

autofocusing

The use of active (e.g. optical autofocusing devices) and passive (e.g. image analysis) methods to maintain a selected plane or degree of focus of a specimen. In scanning electron microscopes, which have a large depth of focus, image analysis software can be used to measure and iteratively correct image sharpness and astigmatism through control of lens currents.

Laser autofocusing device for a light microscope. Courtesy of Prior Scientific Instruments.

autofocusing device

A device that maintains a selected plane of focus in a microscope. In light microscopes, a typical autofocusing device uses an epi-illuminator that introduces a laser beam into the optical pathway; the beam is reflected by the specimen and focused onto a position-sensitive detector. If the focal spot moves, the device sends an adjustment signal to the motor-driven microscope focusing mechanism.

automatic gain control

The automatic adjustment of gain so that the output signal remains at a specified level despite fluctuations in the input signal.

autometallography

A technique for enhancing the contrast of metal probes by the addition of other metals in a process analogous to photographic development. In

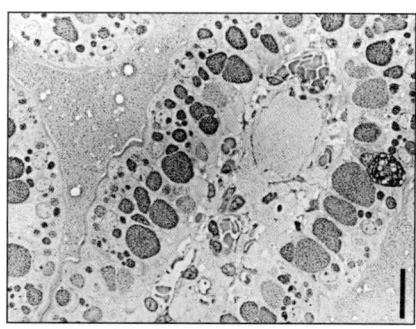

Autometallography. Silver enhanced gold-labeled immunocytochemical localization of pig IgG. Courtesy of Julian Heath and Laszlo Komuves. Bar = 50 μm.

autometallography, sections containing components labeled with metal probes (e.g. Ag, Au, Bi, Hg, Se, Zn) are incubated with gold or silver ions in the presence of a reducing agent (e.g. hydroquinone) causing the metal ions to bind to the probe.

autoradiography

The localization of radioactively labeled cell components by photographic techniques. In autoradiography radiation from the specimen causes development of silver grains in a film emulsion in contact with the specimen. Autoradiography is no longer widely used because of radiation hazards and its low-resolution compared with immunocytochemistry.

autostereoscopy

A technique for the unaided viewing of a stereoscopic image. Autostereoscopic techniques include holography and lenticular autostereoscopy.

auxiliary lens

1. In a light microscope, any extra lens that can be placed in the optical pathway, for example beneath the condenser for use with low NA objectives, or in front of the primary objective of a stereomicroscope.
2. In an electron microscope, any extra lens in the column that is only used for special applications, such as a Lorentz lens.

avalanche photodiode

A photodiode with built-in gain or avalanche multiplication of charge carriers. An avalanche photodiode has a high reverse bias voltage so the electron-hole pairs generated by each incident photon have enough energy to create additional pairs, which in turn create more, so amplifying the signal.

averaging

The formation of an image or signal by using the the mean of two or more separate images or signals.

axial chromatic aberration

The distance along the optical axis between the focal points of light of different wavelengths.

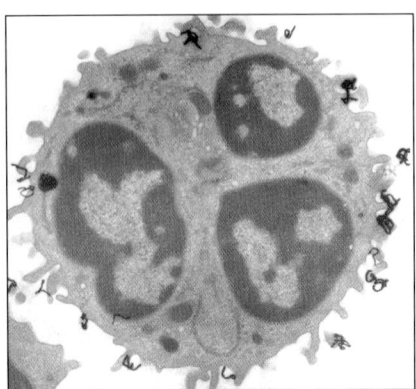

Autoradiograph of radioactively labeled white blood cell. Courtesy of Agar Scientific.

axial magnification

The ratio of the distances between two axial points in the image and the same points in the object.

axial point-spread function

The point-spread function along the optical axis.

axial resolution

In a conventional light microscope the axial resolution r_z of two point objects along the optical axis is given by the expression:

$$r_z = 2\lambda n/\text{NA}^2$$

where r is radius of Airy disk, λ is wavelength, n is refractive index of specimen, and NA is numerical aperture of objective.

axial tomography

Tomography in which the specimen is rotated about its major axis.

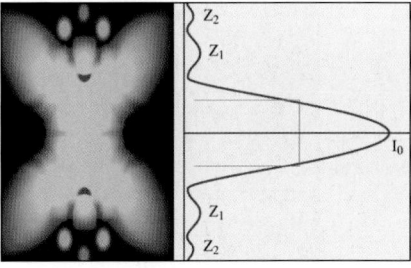

Axial point-spread function of a fluorescence light microscope. Left: vertical section through PSF. Right: intensity distribution along optical axis. I_0: intensity at focal plane; Z_1, Z_2: side maxima. Courtesy of Leica Microsystems.

B

Ray diagram showing back focal plane (bfp) of light microscope objective (O); S, specimen; C condenser. Courtesy of Carl Zeiss Ltd.

Back-illuminated charge-coupled device. Courtesy of Hamamatsu.

Babinet-Soleil compensator

A type of compensator comprising two calcite or quartz wedges placed one on top of the other at the hypotenuse. Sliding one wedge past the other changes the amount of retardation.

back focal length

The focal length on the image side of a lens.

back focal plane

The focal plane on the image side of a lens. In a light microscope the back focal plane of an objective can be viewed with a Bertrand lens or phase telescope to inspect diffraction patterns or align optical elements.

background

Any signal in a detector, image, or spectrum that does not originate from the object or is not a result of the experimental parameters, such as auto-fluorescence or detector noise.

background subtraction

The removal from an image or signal of background contributed by the radiation source, specimen, microscope, detector and other sources.

back-illuminated charge-coupled device

A type of CCD in which the illumination passes through the basal silicon substrate to reach the pixels unimpeded by gate structures. In a back-illuminated CCD the substrate has been thinned by etching, improving sensitivity at all wavelengths.

backing pump

A vacuum pump that receives the gases output by another vacuum pump. In an electron microscope a rotary pump usually backs-up the oil-diffusion pump.

backscattered electron

An incident electron that is inelastically scattered and reflected back towards its source, typically due to multiple elastic and inelastic collisions in the specimen. Backscattered electrons have an energy greater than 50 eV.

backscattered electron detector

A detector for backscattered electrons in scanning and transmission electron microscopes. Two common types of backscattered electron detector are the solid-state and the Robinson detectors.

backscattered electron detector, annular

An annular semiconductor detector for backscattered electrons, widely used in scanning and transmission electron microscopes. The detector is split into four quadrants to allow mathematical operations on the signals: addition of the signals from all four sectors gives a compositional or Z-contrast image; subtraction of alternate sectors gives a topographic image.

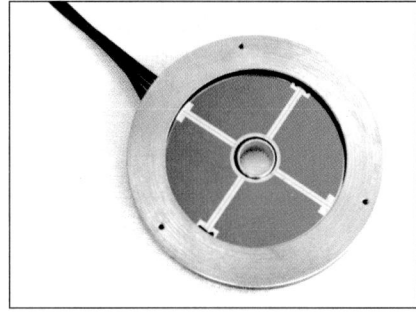

Backscattered electron detector. Courtesy of Agar Scientific.

backscattered electron diffraction see electron backscatter diffraction

backscattered electron imaging

The formation of an image using backscattered electrons, typically performed in a scanning electron microscope. Electron backscattering is strongly dependent upon atomic number, so backscattered electron images can provide compositional information or Z contrast.

backscattered Kikuchi diffraction see electron backscatter diffraction

backscattering

Scattering at an angle greater than 90° with respect to the incident beam.

backthinned charge-coupled device see back-illuminated charge-coupled device

bake out

The removal by heating of gaseous and volatile materials.

balanced-break method

A standard method used to prepare glass knives from glass strips and squares. By balancing the glass on a fulcrum and applying equal pressure on either side, identical squares or triangles of glass can be produced.

ballistic electron-emission microscopy

A type of scanning tunneling microscopy that measures and maps the mean free paths of high energy electrons passing from a conductive tip through the sample to a back electrode.

band gap

The energy required to move an electron from the valence band into the conduction band of a solid.

bandpass

The range of wavelengths or frequencies transmitted by an optical or electronic filter.

bandpass filter

A filter that transmits only a specific range of wavelengths.

bandwidth

The range of frequencies, wavelengths or information in a signal, e.g. the wavelengths of light transmitted by an interference filter.

bar

A non-SI unit of pressure, equal to 100,000 Pa.

barn

A unit of area used to quantify cross-sections in atomic, electron and particle interactions. Symbol b. One barn equals 10^{-28} m^2.

barrel distortion

A type of distortion where the image shows decreasing magnification from center to margins; a square object will appear barrel shaped.

barrier filter

A generic term for any optical filter that blocks a specific range of wavelengths of light, such as shortpass and bandpass filters.

Bayer filter

A mosaic of primary color optical filters placed in front of a CCD in order to create a color image. Each set of adjacent pixels is covered with either

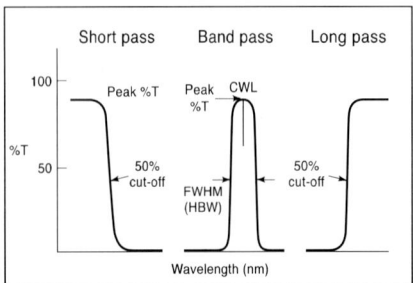

Principle of bandpass filter. From D. Murphy © John Wiley & Sons, Inc.

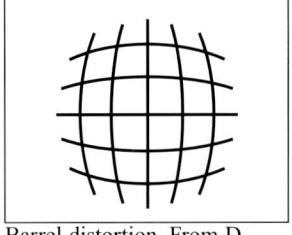

Barrel distortion. From D. Murphy © John Wiley & Sons, Inc.

red, green or blue filters as part of a repeating pattern that covers the whole CCD. Typically there are more green than red and blue filters.

beam

A collective term for a stream of radiation passing through a system.

beam blank

A set of deflection coils that deflect the electron beam from the optical axis of an electron microscope.

beam broadening

The spreading of a beam of radiation. The broadening of an electron beam in a specimen is caused by scattering; its magnitude is a function of beam energy, specimen thickness and specimen density.

beam coils

A collective term for any beam-deflection coils in an electron microscope.

beam current

The current in amperes flowing through the column of an electron microscope, produced by the electron beam. Beam current at the specimen can be measured using a Faraday cup; beam current at the viewing screen is measured for photography.

beam damage see radiation damage

beam deceleration

The deceleration of an electron beam by application of a negative voltage to a monochromator, lens or the specimen. Beam deceleration is used in some electron microscopes to produce a probe with a narrower energy spread.

beam deflectors see deflection coils

beam shift

The lateral displacement of an electron beam accomplished by using the beam shift or beam tilt coils. Beam shifts are used for alignment, for photographic exposures, and as an aid in focusing beam-sensitive specimens.

A standard type of Bayer filter. Courtesy of Julian Heath.

beam spreading see **beam broadening**

beam stop

An opaque object in the path of a beam. A beam stop is used in electron diffraction to block the direct beam.

beam tilt

The angular displacement of an electron beam. Beam tilt is used in centered darkfield, diffraction and stereoscopic modes.

beam waist

The smallest diameter of a focused beam.

beamsplitter

A device that splits transmitted illumination into two or more beams.

beamsplitting prism

A beamsplitter made by cementing two prisms together. The transmitted beams may be non-polarized or polarized. A polarizing beamsplitting prism outputs two parallel or diverging orthogonally polarized beams (the ordinary and extraordinary beams).

Beem capsule

Proprietary type of polyethylene embedding mold.

Beem specimen embedding molds for transmission electron microscopy. Courtesy of Agar Scientific.

bell jar

A thick glass dome in which a vacuum can be formed. A bell jar is usually surrounded by a steel mesh or plastic cage for safety in case of implosion.

bending

The relative values of the curvatures of the two refracting surfaces of a lens.

Bertrand lens

A convergent lens that can be inserted into the tube of a light microscope to form an image of the back focal plane of the objective in the intermediate image plane. A Bertrand lens is used to inspect diffraction patterns, interference images and align optical devices such as diaphragms, phase annuli and prisms.

beryllium window

An X-ray spectrometer window made of a thin (~7 μm) film of beryllium. A beryllium window is suitable for microanalysis of high atomic number elements, but not for elements with Z below 11.

Bessel function

A mathematical function used in the analysis of rotationally symmetric systems.

Bethe-Heitler model

An expression describing the bremsstrahlung X-ray intensity as a function of X-ray energy and specimen atomic number, specifically derived for thin foils and high-energy (>100 keV) electrons typical of analytical electron microscopes. The expression is most useful for modeling the rapid changes in bremsstrahlung intensity in the low-energy (<2 keV) region of the spectrum where absorption of bremsstrahlung X-rays occurs both within the specimen and the energy-dispersive detector.

bias resistor

A variable resistor in the electrical circuit of an electron gun that connects the high voltage supply to the filament.

bias retardation

An additional phase shift between the ordinary and extraordinary waves added by the objective prism in a DIC microscope to increase the intensity of the background in the image.

bias voltage see gun bias

biaxial crystal

A crystal having two optic axes and three indices of refraction. Biaxial crystals are members of the crystallographic groups orthorhombic, monoclinic and triclinic.

biconcave lens

A diverging lens having two concave surfaces.

biconvex lens

A converging lens having two convex surfaces.

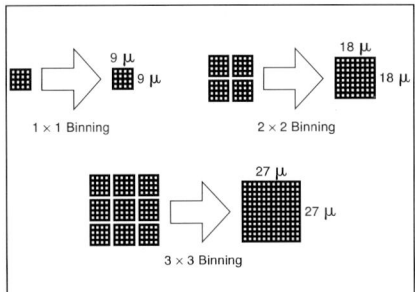

Principle of pixel-value binning and effect on resolution. From D. Murphy © John Wiley & Sons, Inc.

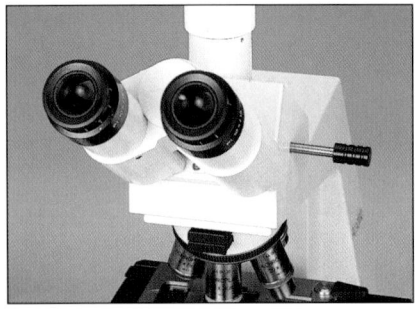

Binocular viewing head on light microscope. Courtesy of Carl Zeiss Ltd.

binning
The addition of the values of adjacent points in an image or signal to produce a single value. In CCD cameras, binning of four (2x2), nine (3x3) or sixteen (4x4) pixels is used for enhancement of low-intensity specimens and for faster imaging, but at the expense of resolution.

binocular
Requiring the use of both eyes.

binocular microscope
A light microscope having two eyepieces. Binocular viewing reduces eyestrain in standard light microscopes by providing each eye with the same image, and allows stereoscopic vision in stereomicroscopes, in which each eye receives a different image.

binocular stereoscope
A stereoscope with two simple eyepieces, presenting one image of a stereo pair to each eye.

binocular tube see binocular viewing head

binocular viewing head
A viewing head on a light microscope that carries two eyepieces.

binocular vision
Seeing with both eyes.

bioluminescence
Luminescence produced by living organisms. The bioluminescence of jelly fish is due to the enzyme aequorin which activates a green fluorescent protein causing it to release light; firefly bioluminescence results from the action of the enzyme luciferase on luciferin.

bioluminescence resonance energy transfer
A form of fluorescence resonance energy transfer that occurs naturally in luminescent organisms, in which the donor is a bioluminescent molecule, often activated enzymatically.

biprism

1. An optical device consisting of two apposed glass prisms, used to deflect, invert, split or polarize a beam.
2. A thin wire used as an electron beamsplitter.

birefringence

1. A widely used synonym for double refraction.
2. The difference between the refractive indices of a double refracting material. Symbol Δn.

bit depth

The number of bits used to digitally encode a signal or pixel. In digital images eight bits allows 2^8 or 256 possible gray or color levels per pixel, 16 bits allows 2^{16} or 65,536 levels (high color) and 24 bits allows 2^{24} or 16,777,216 levels (true color).

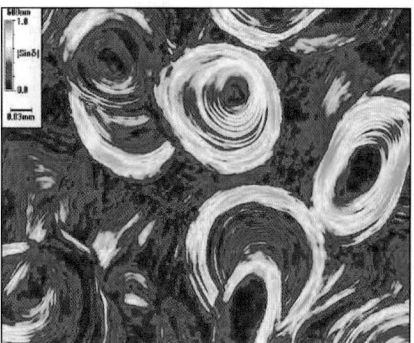
Birefringence. Orientation of slow axis in bone. *Microscopy and Analysis* 2003.

bitmapped image

A digital image that is stored and displayed as a two-dimensional array of pixels.

black and white

A colloquial term for a grayscale image or an imaging device that produces a grayscale image.

black level see offset

bleed through

The presence of spurious signals in a fluorescence image. Bleed through is caused by the overlap of the passbands bandwidth of excitation and emission filters, or, in multiply labeled specimens, the overlap of fluorophore excitation spectra, or excitation of a fluorophore by light emitted from another.

blind deconvolution

A form of image deconvolution in which out-of-focus information is removed without firsthand knowledge of the point-spread function of the microscope. In blind deconvolution the point-spread function is estimated from data within the image; the estimated PSF is then used to remove out-of-focus information in an iterative deconvolution process.

blind spot

The light-insensitive region of the retina where neurons meet to form the optic nerve.

block face

The surface from which sections are cut from a specimen block. For histology the block face is usually rectangular; for ultramicrotomy it is usually a trapezoid.

block see specimen block
block trimming see trimming

blooming

Blooming in a CCD image as light intensity increases. Courtesy of PCO AG.

The broadening of the brightest areas of a video image. Blooming occurs when: excess charge in a CCD pixel well spills into adjacent pixels; a video-tube camera target is saturated; a monitor gain is excessive.

blue fluorescent protein see fluorescent proteins
blur circle see circle of least confusion

BMP image file format

The bitmap image file format used by Windows-based computers. File extension: bmp.

BNC connector

A bayonet-type connector used to couple co-axial cables to video and other electronic equipment.

body tube

That part of the tube of a light microscope which is integral with the stand, typically carrying the nosepiece on one side and the viewing head and additional optical components on the other.

Boersch effect

An anomalous broadening of the energy spread of an electron beam, occurring in regions of high current density.

Brace Köhler compensator

A type of compensator comprising a rotatable thin plate of mica, used for measuring very small retardations.

Bragg condition see **Bragg's law**
Bragg-diffraction contrast see **diffraction contrast**
Bragg equation see **Bragg's law**

Bragg's law

A law prescribing the conditions for constructive interference of waves diffracted by crystal planes. Bragg's law states that two waves reflected from adjacent crystal planes will constructively interfere if the difference in their pathlengths is a whole number of wavelengths. Bragg's law is satisfied when $n\lambda = 2d\sin\theta$, where n is an integer (1, 2, 3, etc.), l is the wavelength of radiation, d is the distance between the two planes, and θ is the angle rays make with the plane.

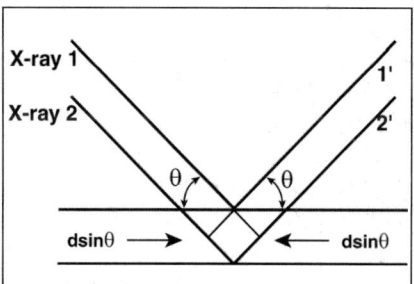

Conditions for Bragg's law. Courtesy of Thermo Electron Corporation.

braking radiation see **bremsstrahlung**

bremsstrahlung

X-rays emitted as a result of collisions between electrons and atomic nuclei. Bremsstrahlung is seen as a background or continuum in an X-ray microanalysis spectrum.

Brewster-angle microscopy

The measurement and mapping of the state of polarization of light reflected from an interface. Brewster angle microscopy is used to investigate the properties of surface films.

Brewster's angle

The angle of incidence of light at an interface that produces maximum linear polarization of the reflected rays. At Brewster's angle θ, the reflected and refracted rays are at right angles, and $\tan\theta$ is equal to the refractive index n of the medium. For water: $n = 1.33$, θ is 53°; for glass: $n = 1.52$, θ is 57°.

bridge

A molecule that links a primary probe to another probe. Bridge molecules are commonly used in cytochemistry for signal amplification.

brightfield microscopy

A mode in light and electron microscopy in which both diffracted and undiffracted (direct) radiation

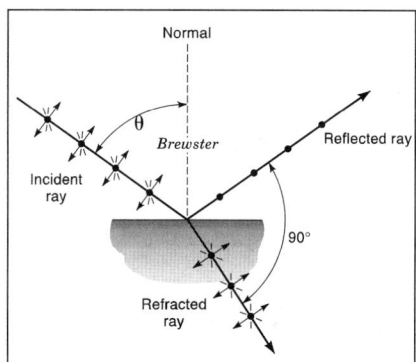

Polarization of reflected light and Brewster's angle. From D. Murphy © John Wiley & Sons, Inc.

from the specimen and its surrounds is collected by the objective lens to form an image of the object on a bright background.

brightness

The qualitative measure of the visual system's perception of the strength of light.

broadband antireflection coating

A multilayer antireflection coating that reduces the reflection of a range of wavelengths of light.

Brownian motion

The random movement of submicrometer-sized particles in a fluid. Brownian motion is caused by collisions between particles and fluid molecules.

buffer tank

A component of a vacuum system that provides a reservoir for the gases evacuated by a vacuum pump. On electron microscopes the oil diffusion pump discharges gases to a buffer tank that is periodically emptied by the backing or rotary pump.

bulk specimen

In electron microscopy, any specimen that is opaque to the electron beam.

burn in

Damage to a phosphor viewing screen as a result of electromagnetic or electron radiation.

burner see arc lamp

C

calcite

Crystalline calcium carbonate ($CaCO_3$). Calcite is optically anisotropic and used in the manufacture of polarizers and retardation plates. RI at 530 nm: $n_e = 1.489$; $n_o = 1.663$.

calibration

The correlation of the measurements obtained from a specimen to those obtained from a calibration standard.

calibration standard

A reference object used for the calibration of microscopes. The object's parameters should be traceable to standards set by a national or international body such as: International Organization for Standardization (ISO); National Institute of Standards and Technology (NIST) in the USA; European Committee for Standardisation (CEN).

calibration standards, for LM

Calibration standards for magnification, resolution and aberrations of light microscopes and cameras include: stage micrometer slides; slides with variably spaced lines, dots and geometric shapes; diatoms; grayscale patterns.

calibration standards, for microanalysis

Calibration standards for microanalysis include samples of elements and compounds with characteristic X-ray lines to calibrate X-ray detector efficiency and resolution. For EDX microanalysis of thin sections, standards include: for materials sciences, salts in the form of dried droplets, or ground-up well characterized minerals; for biological applications, similar standards prepared in a suitable organic matrix such as gelatin.

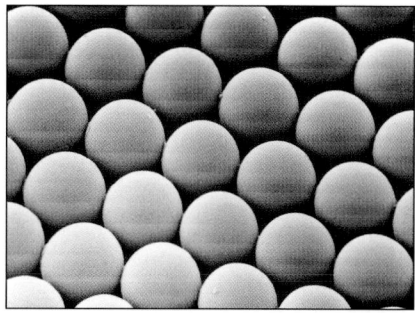

Polystyrene spheres for magnification calibration and astigmatism correction of scanning electron microscopes. Courtesy of Agar Scientific.

calibration standards, for SEM

Standards for the calibration of magnification,

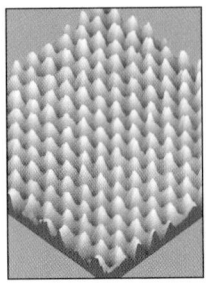

Highly orientated pyrolytic graphite (left) and chess-pattern pillars (right) for magnification and tip calibration of scanning probe microscopes. Courtesy of Agar Scientific.

resolution and aberrations of scanning electron microscopes. Commonly used standards include: particles or gratings for magnification calibration; evaporated metal particles on carbon films for resolution and high magnification calibration; high purity elements for backscattered electron detector calibration.

calibration standards, for SPM

Standards for the calibration of magnification, resolution and aberrations of scanning probe microscopes. Commonly used standards include: gratings and substrates with characteristic topography for magnification, scanner and tip calibration.

calibration standards, for TEM

Standards for the calibration standards of magnification, resolution and aberrations of transmission electron microscopes. Commonly used standards include: diffraction gratings for low magnification calibration; crystals for resolution and high magnification calibration; holey carbon films for astigmatism correction; crystals for diffraction and camera length calibration.

camera

A device for the recording of images on film or a detector.

camera, 35 mm

A camera that uses 35-mm photographic film.

camera, CCD see charge-coupled device

camera chamber

The compartment at the bottom of the column of a transmission electron microscope containing the film cassette or image detector.

camera constant

The product of electron wavelength λ and camera length L in a transmission electron microscope. The camera constant relates the distance R between a diffraction spot and the direct beam to the spacing d of specimen crystal planes: $dR = \lambda L$.

camera length

Notionally the distance from the specimen to the

Diffraction grating for magnification calibration of TEM. Courtesy of Agar Scientific.

screen, film or recording device in a transmission electron microscope in diffraction mode. The camera length increases proportionately with magnification, and can be calibrated using the camera constant and standards with known crystal spacings.

camera lucida

A device to aid the drawing of specimens viewed in a light microscope. A camera lucida is placed in the eyepiece tube and contains a beamsplitting system that allows the image and a sheet of paper beside the microscope to be viewed simultaneously.

camera shutter see shutter

candela

SI base unit of luminous intensity. Symbol cd. One candela is the directional luminous intensity of a source producing light at a frequency of 540 terahertz (555.17 nm) with a power of 1/683 watt per steradian

cantilever

The arm that carries the probe of a scanning probe microscope. The deflection or the changes in resonant frequency or amplitude of the cantilever is the primary source of information and image formation in scanning probe microscopy. Cantilevers are typically made of silicon (Si), silicon dioxide (SiO_2) or silicon nitride (Si_3N_4) and are manufactured by microfabrication techniques such as photolithography and electron-beam or focused ion-beam milling. Standard shapes are the beam and V-shaped cantilevers; the former are used as sensors and when torsional forces are studied; the latter are more stable and used primarily for contact modes. Motion is detected by reflecting a laser beam from an aluminum or gold coated cantilever onto a position-sensitive detector, or by incorporating, or manufacturing, the cantilever from piezoresistive material.

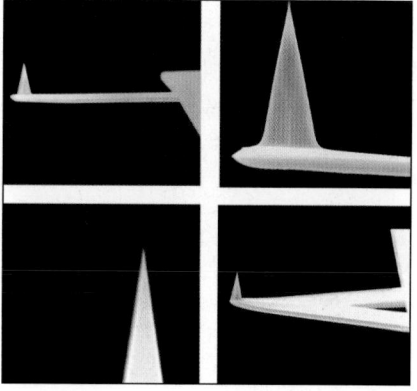

Scanning probe microscope cantilevers and tips. Upper left: beam cantilever; lower right: V-shaped. Courtesy of NT-MDT.

cantilever quality factor see quality factor

cantilever sensor

A cantilever that responds to a physical or

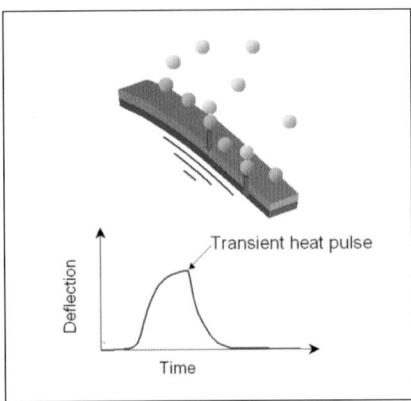

Principle of cantilever sensor. Courtesy of Veeco Metrology.

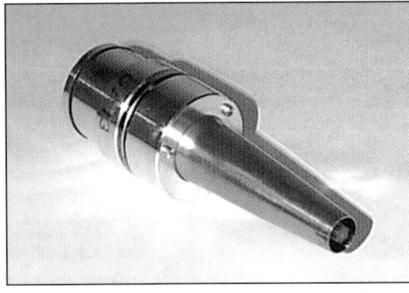

Capillary optic for X-ray microanalysis. Courtesy of Thermo Electron Corporation.

chemical stimulus. Depending upon application, a cantilever sensor may be coated with a material that has selective affinity for a target or manufactured from materials that respond to a physical property of an environment, such as heat. Detection is typically based on the bending of the cantilever arising from added mass, chemical interaction of coating and target molecules, or heating. Cantilever sensors are often manufactured in parallel arrays to optimize detectability.

cantilever sensor array see **cantilever sensor**
cantilever spring constant see **spring constant**
cantilever tips see **tips for SPM**

capillary optic

A lens comprising many micrometer-diameter capillaries, typically used to focus or collimate X-rays. In a capillary optic X-rays propagate along a hollow (curved) capillary tube by multiple grazing-incidence reflections from the internal surface.

capture cross-section see cross-section

carbon braid

A braid of carbon fibers used as an evaporation source in carbon coating.

carbon coating

The coating of a substrate or specimen with a carbon film. Carbon is typically used to provide an electron transparent, conductive coating for electron microscopy.

carbon film

A very thin (<50 nm) coating of carbon, typically used as a support film or to stabilize a plastic support film. Carbon films are widely used in transmission electron microscopy as they are highly electron transparent.

carbon nanotube see nanotube

carbon rod

A rod of carbon or soft graphite used as an evaporation source in a coating unit. Before use, the rods are prepared using a carbon rod shaper to give tips with different profiles, such as cylindrical, angled or stepped.

carbon rod shaper

A device that shapes carbon rods for use in a coating unit.

cardinal distances

The reference distances parallel to the optical axis that define the properties of lenses and optical systems. The cardinal distances include: the focal lengths on the object and image side; the object distance; the image distance; the intersection distances on the object and image side; and the hiatus.

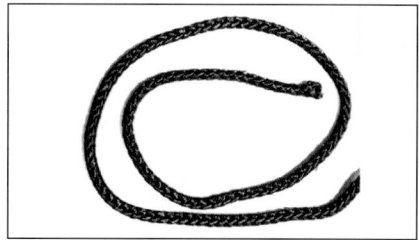

Carbon braid and rods for carbon coating.
Courtesy of Agar Scientific.

cardinal elements

The cardinal distances, planes and points used in geometrical optics to define the properties of a lens or optical system.

cardinal planes

The reference planes normal to the optical axis that define the properties of lenses and optical systems. The cardinal planes include: focal planes, principal planes, nodal planes, object plane and image plane.

cardinal points

The reference points on the optical axis that define the properties of lenses and optical systems. The cardinal points are: the object point, the image point, the focal points on the object and image side; the two principal points; and the two nodal points.

cardioid condenser

A type of achromatic and aplanatic condenser used for darkfield light microscopy. A cardioid condenser contains two mirrors, the lower convex mirror reflects incoming light onto an upper cardioid mirror which directs an oblique cone of light onto the specimen.

cardioid condenser

A condenser for darkfield light microscopy. A cardioid condenser receives light from an annular diaphragm which is then reflected by a central convex mirror onto a outer mirror that projects the light at an oblique angle onto the specimen.

carrier wave

A high frequency wave that is amplitude modulated by another wave signal.

Castaing-Henry filter

An in-column energy filter or electron spectrometer used in transmission electron microscopes. In a Castaing-Henry filter the beam is deflected 90° by a magnetic sector prism, then reflected by an electrostatic mirror and deflected back onto the optic axis by a second 90° magnetic sector prism.

catadioptric

Regarding an optical system containing reflective and refractive elements, such as a mirror objective or Schmidt telescope.

cathode

The more negatively charged terminal of an electrical device, such as the filament of an electron gun and the target in a sputter coater.

cathode ray tube

The device that produces the image in a television or monitor. A cathode ray tube is an evacuated glass tube with an electron gun at one end and a phosphor display screen at the other. A beam of electrons is focused onto the screen and deflected by horizontal and vertical electrostatic plates to make a raster scan producing an image on the screen. The intensity of the scanning beam is determined by changing the bias voltage in the gun. In an SEM the scanning of the electron beam and the CRT are synchronized and the CRT voltage is derived from the electron detector, so the CRT forms an image of the specimen.

cathodic etching see sputtering

cathodoluminescence

Luminescence produced by electrons. Cathodo-luminescence occurs in solids when a valence band electron is excited to the conduction band creating a vacancy that is filled by movement of an electron from the conduction band with the emission of light. The CL emission spectrum, typically covering the range 200-900 nm, is used to characterize the elemental composition of specimens, particularly semiconductors, minerals and rocks.

cathodoluminescence microscopy

The production of an image or map of the cathodoluminescence emitted by a specimen. Cathodoluminescence microscopy can be performed on a light microscope or in a scanning electron microscope. For cathodoluminescence light microscopy a thin section of the specimen is mounted inside a vacuum stage containing an electron gun that irradiates the specimen with 1-30 keV electrons. CL emitted by the specimen passes through a quartz window and can be directly viewed through the microscope objectives and analyzed using a CL spectrometer. In an SEM, CL from the specimen is collected by a light guide and passed through the specimen chamber wall to the spectrometer.

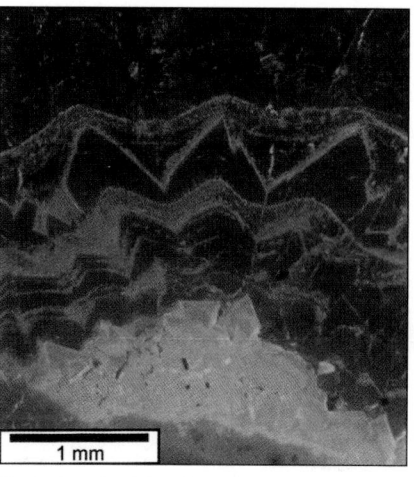

Optical cathodoluminescence of calcite. *Microscopy and Analysis* 2000.

cathodoluminescence spectrometer

A spectrometer used for the analysis of cathodoluminescence spectra. In a CL spectrometer the light emitted from a CL imaging system is dispersed by a diffraction grating and the intensities of each wavelength are measured with a CCD or photomultiplier tube.

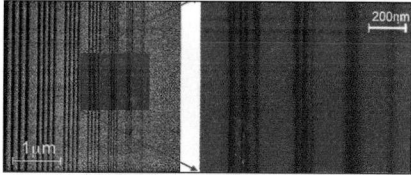

SEM cathodoluminescence of semiconductor. *Microscopy and Analysis* 2002.

catoptric

Regarding an optical system containing reflective elements, such as some astronomical telescopes.

caustic figure

An image formed by the intersection of reflected or refracted parallel rays from a curved surface under conditions of spherical aberration. In electron microscopy, a caustic figure is used to align the instrument.

CCIR video standard

The Consultative Committee for International Radio standard video signal format of 625 lines, 2:1 interlaced, 25 frames per second used in Europe and other parts of the world.

cell culture microscope see **tissue culture microscope**
cement see **optical cement**

center wavelength
The midpoint of the range of wavelengths transmitted by a filter.

centered darkfield
A mode of darkfield imaging in a transmission electron microscope in which the incident beam is tilted so that scattered electrons lie on the optical axis and the objective diaphragm excludes unscattered electrons.

centering stage
A rotating stage with a eucentric mechanism to allow the axis of rotation of the specimen to be centered on the optical axis.

centering telescope
A small telescope that can be placed in the eyepiece tube of a light microscope and used to examine the back focal plane of a light microscope objective in order to align condenser diaphragms, phase annuli or to examine diffraction patterns.

Centering or phase telescope.
Courtesy of Carl Zeiss Ltd.

chamber see **specimen chamber**
channeling see **electron channeling**
characteristic length see **extinction distance**

characteristic peaks
Peaks in an X-ray microanalysis spectrum that are contributed by an ionization event in a specific inner shell of an element.

characteristic X-rays
X-rays with energies that are characteristic of a ionization event in a specific atom.

charge-collection microscopy see **electron beam-induced current**

charge-contrast imaging
A mode of image formation in a variable pressure or environmental SEM which variations in the trapping of charges on the surface of a non-conductive specimen are used to identify features.

charge-coupled device

A semiconductor-based image detector with an array of linked (coupled) photodiodes that generate charges when irradiated with light, X-rays or electrons. A charge-coupled device is an array of metal oxide semiconductor (MOS) capacitors comprising three layers: a basal p-type silicon substrate, a central silicon dioxide layer and an upper polysilicon electrode or gate; each capacitor (also called a photodiode or pixel) is surrounded by transfer gates and potential barriers. A positive voltage applied at a pixel gate moves positively charged holes into the Si substrate creating a depletion layer or potential well for negative charges at the Si-SiO$_2$ interface. When exposed to radiation each pixel generates photoelectrons that accumulate in its potential well. After an exposure, the electron charge packets in each row of pixels are moved in parallel by transfer gates to the serial shift register, a final row of pixels at the edge of the CCD. The serial register moves its contents to the output node and receives the next row, repeating the process until every row of the CCD has been read out. The charges from the entire CCD pass through the output node serially and are finally amplified, digitized and processed to form the image.

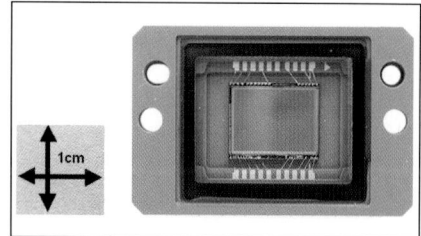

Charge-coupled device. Courtesy of PCO AG.

charge-coupled device camera

A camera containing a charge-coupled device. CCD cameras are used for imaging and detection in all forms of microscopy.

charge-injection device

A solid-state imaging device in which each pixel is individually addressed during readout. After read-out the charges remain in the pixel and are cleared by injection into the substrate.

charge transfer

The process by which the electrons in one potential well of a CCD are moved to an adjacent well.

charge transfer efficiency

The fraction of a charge packet remaining in a well of a CCD pixel during charge transfer. The charge transfer efficiency affects read-out rates of CCDs.

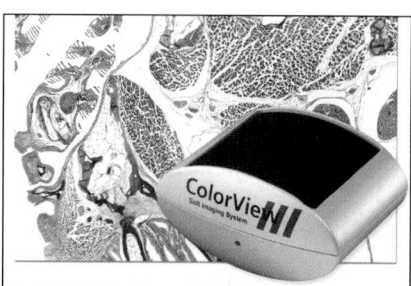

CCD cameras for light microscopy (top) and electron microscopy (bottom). Courtesy of Soft Imaging System.

charged particle

An elementary particle carrying a positive or negative charge, such as an electron or proton.

charging

The build-up of negative charges on a specimen irradiated with an electron beam. Charging may occur in a scanning electron microscope when there is poor electrical conductivity of the specimen.

chatter

A sectioning artifact caused by high frequency vibrations leading to variations in section thickness parallel to the knife edge. Possible causes of chatter include: block face too large; sections too thick; knife angles incorrect; or cutting speed too high.

chemical fixation

The fixation of a specimen using a chemical to cross-link or stabilize molecular structures.

chemical force microscopy

The use of a scanning probe microscope to measure the forces involved in molecular interactions.

In chemical force microscopy the probe is functionalized by addition of a specific molecule or ligand that can interact or bind to a receptor on a substrate or on the specimen. Force versus distance curves are obtained by moving the tip towards the specimen, allowing the ligand to bind to its receptor, and then moving the probe away from the specimen.

Chemical force microscopy: interaction of chemical tip with specimen. *Microscopy and Analysis* 2003.

chemical-state mapping

The use of spectroscopic techniques such as electron energy-loss spectroscopy or X-ray microanalysis to map the chemistry or chemical bonding states of a specimen.

chemical tip

A scanning probe microscope tip which has been chemically treated or to which molecules have been attached.

chief ray

A ray that starts from the edge of the object and passes through the centre of the entrance and exit pupils, defining the height of the image.

chisel-nose detector

An X-ray spectrometer with a detector mounted at an angle, a design used when the axis of the spectrometer arm is parallel to the specimen plane.

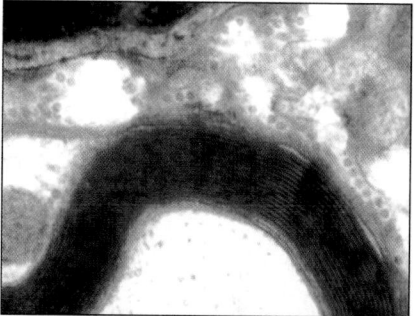

Chromatic aberration. From D. Murphy © John Wiley & Sons, Inc.

chromatic aberration

An aberration of lenses in which focal length varies with wavelength as a result of dispersion. Chromatic aberration is a major problem in electron microscopes due to significant energy losses of scattered electrons.

chromatic aberration, in TEM

Chromatic aberration in transmission electron microscopy arises primarily from the energy spread of the gun emission, high-voltage instabilities, and energy losses as the beam passes through the specimen, which are especially prevalent in thick specimens. Chromatic aberration in electron microscopes is of three kinds: chromatic aberration (coefficient C_C), which depends linearly on inclination of the ray to the optic axis at the object and on the relative energy spread and so is most serious for objective lenses; chromatic aberration of magnification, which depends linearly on off-axis distance at the object and on the relative energy spread and so is most serious for projector lenses; and for magnetic lenses there is also an anisotropic chromatic aberration. The radius of the chromatic aberration disk R_{chr} is given by:

$$R_{chr} = C_C . \delta E / E_0 . b$$

where δE is energy loss, E_0 is accelerating voltage, and b is semiangle of collection of the lens. The formula for C_c shows that chromatic aberration cannot be eliminated from round lenses by skillful lens design.

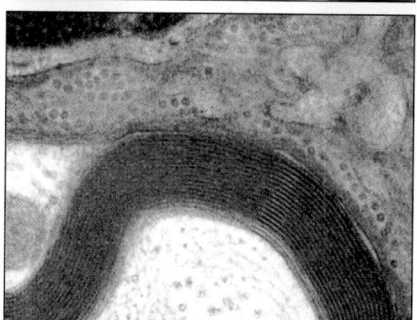

Degradation of resolution of TEM image of 0.5-μm thick section due to chromatic aberration (top). Resolution is improved in energy-filtered image (bottom). Courtesy of Carl Zeiss SMT.

chromatic difference of magnification

The difference in the magnification of images of the same object formed with radiation of different wavelengths.

chromatic dispersion see dispersion

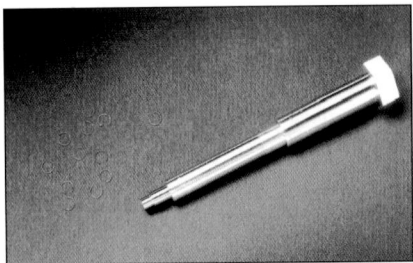

Circlips and circlip injector. Courtesy of Agar Scientific.

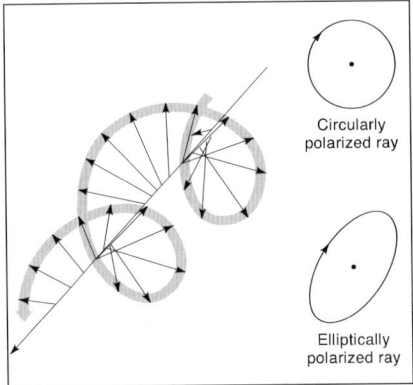

Circularly and elliptically polarized light. From D. Murphy © John Wiley and Sons, Inc.

chromaticity

The quality of the color of light as defined by its dominant wavelength and purity.

chrominance

The color and saturation information in a video signal.

chromophore

A chemical domain that imparts color to a stain.

chuck see microtome chuck

ciliary muscles

A circular set of muscles that surrounds the lens of the eye. When the ciliary muscles contract tension in the suspensory ligament that connects the lens and ciliary muscles is reduced increasing curvature of the lens; ciliary muscle relaxation increases tension in the suspensory ligament, stretching the lens and reducing its curvature.

circle of least confusion

The smallest diameter spot formed in an image of a point object by an optical system that is not fully corrected for spherical aberration, chromatic aberration or astigmatism.

circlip

A C-shaped metal clip used for retaining support grids in an electron microscope specimen holder.

circlip injector

A tool for placing circlips into a specimen holder.

circular stage see rotating stage

circularly polarized light

Light waves in which the electric field vector rotates 360° in each wavelength, describing a helix about the axis of propagation. Circularly polarized light is produced when two linearly polarized waves with a relative phase shift of $n\lambda/4$ combine where n is an integer.

clamping device

A spring-loaded ring that clamps an EM grid in place in the specimen holder.

clearance angle

The angle between the block face and the front surface of a microtome knife. The ideal clearance angle for sectioning is about 5°.

Cliff-Lorimer ratio

A method for quantifying the elemental information in X-ray microanalysis spectra from thin specimens. The relative concentrations C_A, C_B of two elements A and B are directly related to their characteristic X-ray intensities (above background) I_A, I_B through the Cliff-Lorimer sensitivity factor k_{AB} by the relationship:
$$C_A/C_B = k_{AB} \cdot I_A/I_B.$$
The k factor is a function of the atomic number of the elements and can be modified to account for absorption and fluorescence effects if necessary.

closed-loop scanning

The monitoring and correction of the non-linear behavior of the scanner in a scanning probe microscope. In closed-loop scanning sensors in the scanner detect the actual position of the scanner and compare it to the intended position; a feedback loop corrects the voltage applied to the scanner to return it to the intended position. Closed-loop scanning systems typically correct non-linearity, hysteresis and creep of the scanner and allow zooming from a large to a small scanned area of the specimen.

C-mount

A standard type of screw-threaded mounting used to connect a camera to a light microscope. A C- mount ensures optimal positioning of the camera in the optical train.

CMYK

The subtractive color system used in the four-color process by the printing industry. Colors are created by mixing cyan, magenta, yellow, and black (K for key, to avoid confusion with B in RGB).

coat see coating

coating

1. A layer of material applied to a specimen, substrate or optical device.
2. The process by which a coat is applied.

coating device see **coating unit**

coating unit
An apparatus for the coating of specimens and substrates and specimens by evaporation or sputtering of a target in a vacuum.

coherence
The ability of a beam to interfere with itself, i.e. the waves are monochromatic, have a constant phase relationship, and identical state of polarization.

coherence length see **temporal coherence**

coherence probe microscopy
The use of white light interferometry for the measurement of surface relief in a reflected light microscope.

coherence width see **spatial coherence**

coherent anti-Stokes Raman microscopy
A type of Raman microscopy that detects anti-Stokes Raman scattered light. In coherent anti-Stokes Raman microscopy, two near-infrared picosecond laser beams, a pump beam ω_P and a Stokes beam ω_S, are focused with a high NA objective onto the specimen inducing molecular vibrations, leading to the release of a photon of energy $2\omega_P - \omega_S$. The specimen is scanned through the beam and the anti-Stokes Raman signals are collected by a second objective and analyzed by a spectrometer.

coherent bremsstrahlung
Sharp peaks in the bremsstrahlung continuum in an X-ray spectrum generated by similar excitation events in a crystalline lattice.

coherent scattering
The maintenance of coherence following elastic scattering.

coherent source
A source that emits coherent radiation. Monochromatic radiation from a point source is said to be coherent, and the amplitude of the corres-

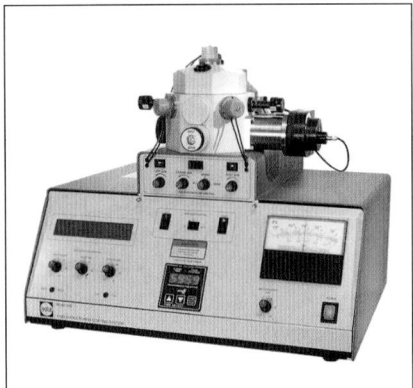

Ion-beam coating unit. Courtesy of Gatan, Inc.

ponding wave at any point is constant while the phase varies linearly with time. In reality, no source provides truly monochromatic radiation and sources are never true points. Real sources emit partially coherent radiation. The wavelength spread is associated with temporal partial coherence and the finite source-size with spatial partial coherence. In electron optics, the contrast-transfer functions are modulated by damping functions, representing the effects of temporal and spatial partial coherence.

cold field emission
Field emission at ambient temperature.

cold field-emission gun
A field-emission gun that uses a cold field-emitter to produce the electron beam.

cold field emitter
A filament that releases electrons by the cold field-emission mechanism.

cold finger
An anticontaminator comprising two metal blades with apertures that are positioned above and below the specimen holder in a transmission electron microscope. The blades are cooled by liquid nitrogen providing a sink for the condensation of contaminants.

cold stage
A microscope stage that is cooled by the Peltier effect or by a cryogen such as liquid nitrogen.

cold trap
An anticontaminator that is cooled below ambient temperature to provide a surface for the condensation of contaminants.

cold-cathode gauge
An ionization pressure gauge in which the cathode is not heated.

collection angle
1. The solid angle subtended at a point in the specimen by the active area of the detector of an X-ray spectrometer.
2. The semiangle subtended at a point in the

specimen by the margins of the entrance aperture of an electron energy-loss spectrometer.

collection efficiency

The efficiency with which a detector collects or receives all the signals emitted from the specimen. In X-ray microanalysis the collection efficiency of a given X-ray is determined by the solid angle of the detector and by the composition of the detector window. Collection efficiency is not the same as quantum or detection efficiency, which is a property of the detector itself.

collection mode

In near-field scanning optical microscopy, the use of the probe to collect light emitted or scattered by the specimen, which is illuminated by an external light source.

collector lens

A lens that projects an image of a light source into an optical device.

collimate

To make parallel a beam of radiation.

collimator

A device containing an aperture that collimates a beam of radiation.

collodion

A solution of nitrocellulose in organic solvent used to produce plastic TEM support films.

colloidal gold

A dense particulate probe widely used in immunocytochemistry. Colloidal gold particles are typically 5-40 nm in diameter and are capable of electrostatically adsorbing a wide variety of macromolecular cytochemical probes such as antibodies, lectins and enzymes. For light microscopy applications colloidal gold particles can be enhanced by autometallography; for SEM and TEM applications, the particles are easily detected of their high electron scattering and Z contrast.

color

Color is not a property of light but a complex

physiological and psychological response of the eye and brain to stimulation of the retina by one or more wavelengths in the visible spectrum.

color CCD camera

A CCD camera used for color imaging. A CCD does not recognise color so a variety of strategies may be used for color imaging. Single-CCD cameras may have a Bayer filter in front of the chip, or use a filter wheel or a tunable acousto-optical beamsplitter to sequentially acquire red, green and blue images that can be merged to form the full color image. Three-CCD cameras may use a combination of prisms and filters to acquire red, green and blue images simultaneously. An alternative technology is to vertically separate the charges in the CCD photodiodes since blue light is absorbed in the surface layer, green in intermediate and red penetrates to the bottom of the photodiode.

color compensation

The adjustment of the source heating current or the use of color filters in order to match the emission spectrum of a light source to the color temperature of photographic color film.

color contrast

The perception by the eye and brain of differences in chromaticity.

color correction, color conversion see color compensation

color filter

A filter that transmits light of one colour or a specific range of wavelengths.

color temperature

A measure in degrees Kelvin of the color spectrum emitted by a light source. Typical daylight has a color temperature of ~5500K; a tungsten lamp filament has a color temperature of ~3200K. Film manufacturers use color temperature to optimize the sensitivity of color film. If the color temperature of the light source is too low, images have a yellow tint; if the color temperature is too high, there is a blue tint.

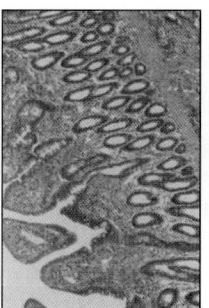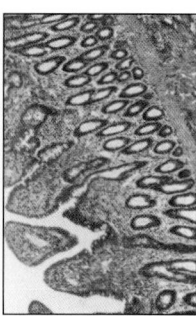

Color compensation. Ektachrome images taken with light source at 3200K without (left) and with (right) color correction filter. Courtesy of Julian Heath.

Column of transmission electron microscope. Courtesy of FEI Company.

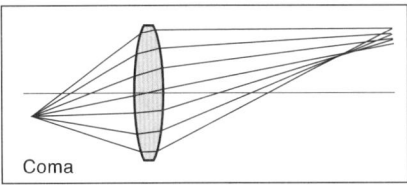

Coma

Coma. From D. Murphy © John Wiley and
Sons, Inc.

color tint plate see full-wave plate

column, of EM

The tall cylindrical region of an electron
microscope that is formed by the stack of round
magnetic lenses between the gun and the
projection chamber. The column contains a central
bore through which the electron beam passes.

column liner see liner tube

coma

An aberration of lenses that causes rays from off-
axis object points passing through the margins of a
lens to be focused in a different plane from rays
passing through the center of the lens. Coma
causes a point object to be comet-shaped with the
tail extending outwards. Coma depends
quadratically on angle and linearly on off-axis
distance at the object; it is the second most
important aberration of magnetic objective lenses
after spherical aberration. Lenses must meet the
sine condition to be free of coma.

common main-objective stereomicroscope

A type of stereomicroscope having a single large
objective lens that is shared by separate parallel
optical pathways leading to each eyepiece. A
common main-objective stereomicroscope usually
has infinity optics allowing the placement in the
tube of accessories such as zoom lenses and epi-
illuminators.

comparison microscope

A light microscope with two optical trains -
illumination source, condenser, stage, and
objectives - linked by a single viewing head that
displays side- by-side images of two specimens in
the eyepiece(s).

compensating eyepiece

An eyepiece that corrects lateral chromatic
aberration and other aberrations in the
intermediate image.

compensator

A retardation plate that is capable of imposing a
variable retardation on transmitted E and O rays.

Compensators are used to measure optical path differences in a specimen by interference and polarized light microscopy. A compensator can be tilted or rotated to produce variable phase shifts from fractions to several wavelengths.

complementary metal-oxide semiconductor

A metal-oxide semiconductor with pairs of n-type and p-type transistors.

complementary metal-oxide semiconductor detector

A semiconductor detector in which each photo-diode (pixel) has a read-out circuit and amplifier.

complementary replica device

A hinged, spring-loaded device that fractures a specimen into two parts to allow the production of complementary replicas.

complementary replicas

Replicas of the two complementary faces produced by a fracture plane through a specimen. Complementary replicas are typically required in the freeze fracture technique to study transmembrane proteins where protrusions in one replica can be matched to complementary pits in the other.

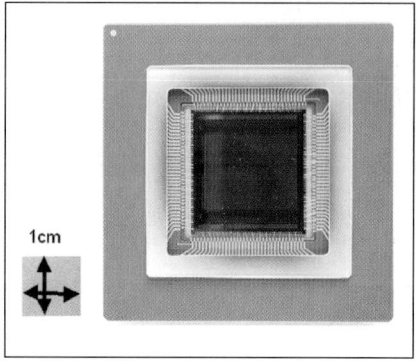

CMOS detector. Courtesy of PCO AG.

complex radiation see polychromatic radiation

composite video signal

A standard video signal containing picture information (luminance, chrominance) together with blanking and synchronization signals.

compositional analysis

The investigation of the presence, concentration, spatial distribution and organization of the structures, molecules, elements or atoms in a specimen by any form of spectroscopy.

compound eyepiece

An eyepiece that contains two lenses or lens systems, the eyelens and the field lens.

compound light microscope

A light microscope having two image-forming optical elements, the objective lens and the eyepiece; the objective produces a magnified intermediate image which is further magnified by the eyepiece to form the final image.

compound objective

A microscope objective with more than one lens.

compression

A sectioning artifact caused by friction between knife edge and specimen. Compression causes a shortening of the section in the dimension perpendicular to the knife edge. Section compression can be minimized by spreading floating sections with heat or solvents.

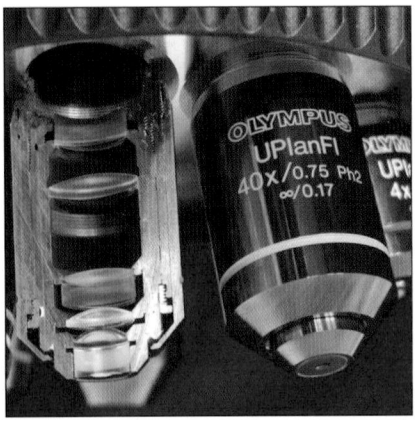

Compound light microscope objective (right) and a view of the multi-element lens system in the cutaway (left). Courtesy of Olympus.

Compton profile

A peak in the high-energy loss region of an electron energy-loss spectrum. The Compton profile arises from high-angle inelastic scattering of electrons and provides data on the ground-state electron momentum distribution.

Compton scattering

The inelastic scattering of electrons and X-rays due to collision with outer-shell electrons.

computed tomography

The use of a computer and digital image processing software in tomography.

computer-aided reconstruction

The use of digital image-processing computer software to obtain the complete structure of a specimen from a set of images: e.g. a complete set of serial thin sections through the specimen, or a set of differently angled projection images obtained by tomographic techniques.

computer-assisted tomography, computerized tomography see computed tomography

condenser

A device that collects and focuses radiation onto a specimen.

condenser, of light microscope

A condenser situated between the light source and the stage of a light microscope that collects and focuses light onto the specimen. The standard features of a condenser are one or more lenses, an adjustable aperture diaphragm, a turret that houses the aperture diaphragm and other optical devices, and a mechanism for moving the condenser axially to focus light onto the specimen.

condenser annulus

An annular diaphragm placed in the front focal plane of the condenser for phase contrast light microscopy.

Light microscope condenser. Courtesy of Carl Zeiss Ltd.

condenser aperture, of EM

The aperture (diaphragm) associated with each condenser lens of an electron microscope. Condenser aperture 1 (Ca1) is usually a fixed aperture located beneath condenser 1 that stops electrons traveling at high angles from the gun crossover point from hitting the column and producing X-rays. Condenser aperture 2 (Ca2) is a changeable aperture (typically 300, 200 or 100 μm diameter) located beneath condenser lens 2. A small Ca2 increases beam coherence.

condenser diaphragm, of LM

An iris diaphragm located in the front focal plane of the condenser that modifies the condenser aperture of a light microscope. The condenser diaphragm controls the angle of rays illuminating the specimen; closing the iris diaphragm reduces image resolution but increases image contrast.

condenser lens, of EM

The magnetic lens that focuses the electron beam onto the specimen. In most electron microscopes there are two condenser lenses, C1 and C2, which gives two degrees of freedom allowing independent selection of the size and angular spread of the beam. In TEMs there may be three condenser lenses: C3 being a mini-condenser lens associated with the objective lens.

condenser lens 1

The first condenser lens in an electron microscope that determines the spot size. C1 has as its object

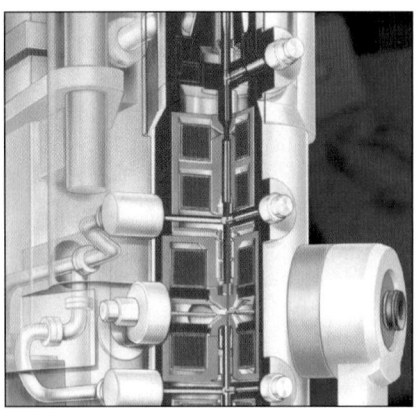

C1 and C2 condenser lenses (top) and objective lens (bottom) of transmission electron microscope. Courtesy of FEI Company.

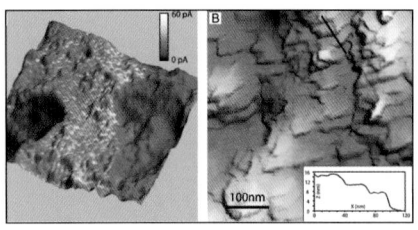

Conductive force AFM image of polymer. *Microscopy and Analysis* 2005.

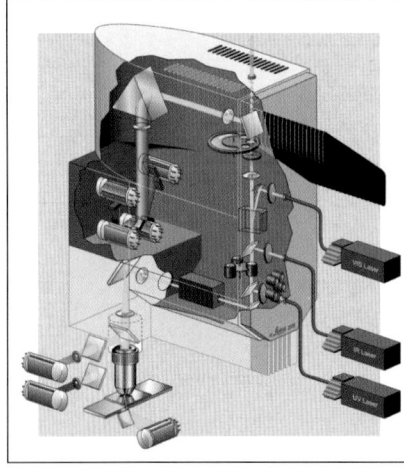

Confocal head: schematic showing optical paths, beamsplitters and detectors. Courtesy of Leica Microsystems.

the crossover of the electron beam. In TEMs with thermionic guns, C1 demagnifies the beam crossover to one of a range of spot sizes, typically from 1-20 μm, selected with the spot size control.

condenser lens 2

The second condenser lens in an electron microscope. In standard parallel-beam mode, C2 projects a defocused image of the gun crossover at the specimen plane. The C2 lens current is set by the brightness or intensity control.

condenser-objective lens

A magnetic objective lens in which the specimen is situated at the centre of the field. The part of the magnetic field upstream from the specimen is thus part of the condenser system, the part downstream is the objective. A condenser-objective lens may be used in parallel-beam and convergent-beam modes.

condenser prism

A Wollaston or Nomarski prism located in the front focal plane of the condenser of DIC light microscope. Condenser prisms are available in different specifications to match specific objectives.

condenser stigmator see stigmator

condenser turret

A rotatable disk in a light microscope condenser that has several openings for diaphragms, phase annuli, prisms and filters.

conductive adhesive

An adhesive containing metal particles suitable for attaching specimens to a substrate where good electrical conductivity is required.

conductive atomic force microscopy

A mode of atomic force microscopy in which the current flowing from a dc-biased tip is used to generate an image of a conductive specimen.

conductive paint

A conductive adhesive used to attach specimens to substrates. Comprises a colloidal suspension of gold, silver or graphite in water or organic solvent to ensure good electrical conductivity between

specimen and substrate

cone cell

A photoreceptor cell in the retina involved in color vision and visual acuity. Cones contain one of three varieties of the visual pigment iodopsin, sensitive to either red, green or blue light. The brain analyzes the relative proportions of the signals from cone cells and interprets them as specific colors.

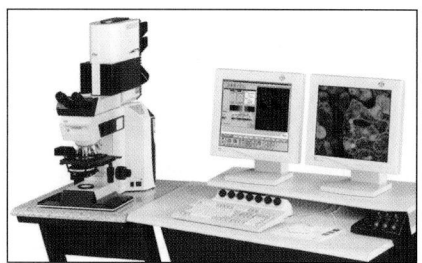

Confocal laser scanning microscope.
Courtesy of Leica Microsystems.

confocal

Having conjugate focal planes.

confocal disk scanning unit

A confocal head with a spinning pattern of slit apertures that create a virtual pinhole.

confocal head

The module of a confocal microscope that contains the optical and mechanical devices and detectors for scanning the specimen and receiving signals.

confocal laser scanning microscope

A type of confocal microscope in which the specimen is illuminated raster-fashion with a focused laser beam; signals emitted by the specimen are collected by a photomultiplier detector one pixel at a time to build an image.

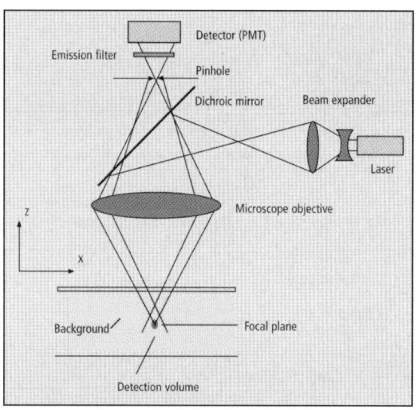

Principle of confocal microscopy.
Courtesy of Carl Zeiss Ltd.

confocal laser scanning microscopy

The use of a confocal laser scanning microscope.

confocal microscope

A light microscope in which a point source, an illuminated volume in the specimen and a small aperture in front of a detector are in conjugate planes. This optical arrangement prevents light from out-of-focus regions of the specimen contributing to image formation, producing clear optical sections of thick and fluorescently labeled specimens.

confocal microscopy

The use of a confocal microscope.

confocal Raman microscopy

The use of confocal optics in Raman microscopy and spectroscopy.

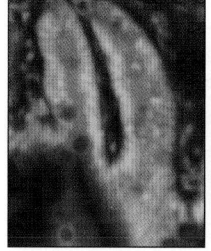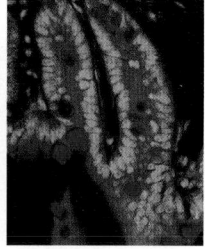

Confocal fluorescence microscopy. Non-confocal (left) and confocal (right) imaging of triple-labeled mouse intestine section.
Courtesy of Carl Zeiss Ltd.

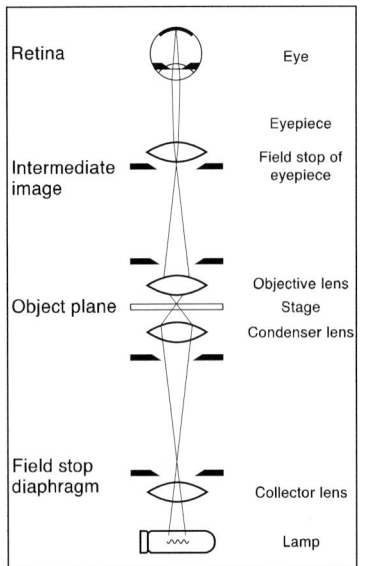

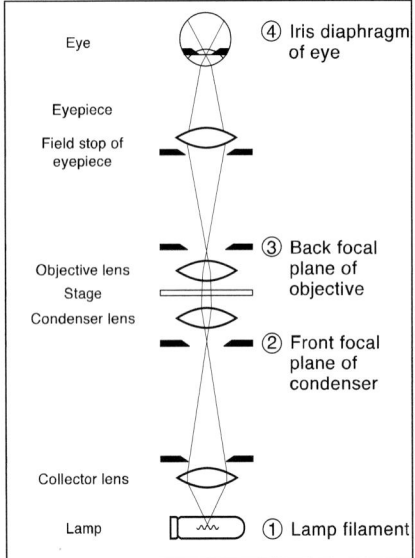

Conjugate field planes (top) and aperture planes (bottom) of a light microscope. From D. Murphy © John Wiley and Sons, Inc.

confocal scanner/scanning unit see **confocal head**
confocal scanning laser microscope see **confocal laser scanning microscope**
confocal tandem scanning microscope see **tandem scanning confocal microscope**

confocal theta microscopy

A mode in confocal fluorescence microscopy that uses two differently orientated objectives, tilted by an angle θ (typically 90°), to enhance resolution; one objective illuminates the specimen and the other detects it. Confocal theta microscopy produces an image with an isotropic point spread function by using a second objective focused on the mid-point of the axially elongated or elliptical PSF created by the illuminating objective.

conjugate

Optically linked.

conjugate planes

A set of planes in an image-forming system that are optically linked and whose images are superimposed at the detector or eye of the observer.

conjugate planes, in light microscope

In a light microscope there are two sets of conjugate planes: aperture planes and field planes. The conjugate aperture planes are: 1. the lamp filament; 2. the iris diaphragm of the condenser; 3. the back focal plane of the objective lens; and 4. the exit pupil of the eyepiece, which is coincident with the pupil of the eye. The conjugate field planes are: 1. the field iris diaphragm; 2. the specimen plane; 3. the intermediate image plane; and 4. the retina of the eye, or the surface of the detector.

conjugate planes, in TEM

In a transmission electron microscope there are two important sets of conjugate planes: 1. the object plane, the intermediate image plane and the image plane; 2. the beam crossover, the back focal plane of the objective (for parallel illumination), the intermediate diffraction plane(s) and the image plane (when the lens settings for observation of the diffraction pattern have been selected). In the STEM, the two important sets of conjugate planes are: 1. the source, the selected-area aperture and the specimen; 2. the virtual objective aperture, the real objective aperture (pivot point) and the dark-field detector.

conoscopic mode

The observation of the back focal plane of a light microscope objective with a Bertrand lens or phase telescope, using a cone of light from the condenser.

constant amplitude mode

A mode in scanning probe microscopy in which the amplitude of an oscillating cantilever or probe is kept constant.

constant current mode, in STM

The operation of a scanning tunneling microscope with a constant tunneling current between the tip and the specimen. As the tip moves over features with different topography or electronic properties, the tunneling current will vary; a feedback loop to the scanner then adjusts the height of the specimen restoring the tunneling current to the constant value. The signals sent to the scanner provide the dataset for the formation of an image.

constant force mode

The operation of an atomic force microscope with constant deflection of the cantilever. As the cantilever bends further due to attraction between tip and a feature on the specimen surface, the scanner moves the specimen (or cantilever) away from the specimen, maintaining a constant cantilever deflection. The feedback signal is used to form the image.

constant height mode, in AFM

The operation of an atomic force microscope with a constant height of the cantilever and probe above the specimen. As the cantilever bends due to attraction between the tip and a feature on the specimen surface, the cantilever motion sensor records the deflections and uses these signals to generate the image.

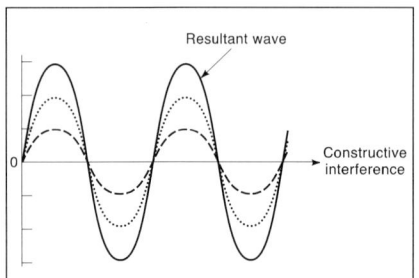

Constructive interference of waves. From
D. Murphy © John Wiley and Sons, Inc.

constant height mode, in STM

The operation of a scanning tunneling microscope
with the tip in a fixed plane above the specimen.
As the tip moves over specimen features with
different topography or electronic properties, the
tunneling current varies; the recorded current is
used to form the image.

constructive interference

Interference in which the resultant wave has an
amplitude greater than that of the interacting
waves.

contact freezing

Cryofixation by contact with a cold surface, such
as freezing pliers, or a metal mirror.

contact mode

The operation of a scanning probe microscope
with the tip and sample in close contact, i.e. within
the repulsion zone of the intermolecular force
curve.

contact profilometer

A profilometer with a sharp steel or diamond
stylus that is in contact with the specimen. The
stylus is coupled to a piezoelectric or optical
sensor that detects and measures vertical stylus
motion during a line or raster scan.

contamination

The adsorption of molecules, particularly
decomposed carbon, by the specimen and
instrument components. Contamination is an ever-
present problem in electron microscopes and leads
to reduction of instrument resolving power, image
resolution and filament life. Primary sources of
contamination are: the specimen itself, which can
release unstabilized components or be sputtered or
vaporized by the electron beam; the vacuum
system, e.g. cracked hydrocarbons from diffusion
pump oil; water vapor from the specimen or
photographic film; and vapor leaks from O-rings.
Solutions include: cleanliness in handling of
microscope parts; cleaning specimens and
instrument parts with a plasma cleaner; outgassing
of photographic films; use of anticontaminators;
and maintenance of high vacuum conditions.

continuous dynode

A glass vacuum pipe coated with an electron emitting substance. As electrons flow through the pipe under an applied voltage they collide with the pipe walls producing secondary electrons that in turn generate more electrons, amplifying the original signal by several orders of magnitude.

continuum radiation see bremsstrahlung

contrast

The difference in absolute or perceived intensity between an object and its surroundings. Contrast may be physiological or photometric.

contrast, in TEM

In a transmission electron microscope all incident electrons emerge from the specimen and so there is no real absorption contrast. Electrons are scattered by the atoms in the specimen and these deviations are transformed into contrast variations at the image by various mechanisms. At relatively low resolution, contrast is created by the interception of scattered electrons by the objective lens aperture (diaphragm). Electrons that have passed close to heavy atoms are thus absent from the image and a form of amplitude contrast results. At higher resolutions, contrast is created by interference between the unscattered wave (corresponding to electrons that have passed through the specimen undeviated) and scattered waves. The combined effect of spherical aberration and a deliberate defocus alter the phase difference between these waves and contrast is again created in the image. This behavior is characterized by the phase contrast-transfer function. There is no true amplitude contrast but such contrast does occur as a result of elastic scattering and of departures from the assumption that the phase variations are small.

contrast transfer function

A measure of how faithfully contrast in an alternating black and white (square wave) object is transferred by an optical system to the image. When object contrast varies sinusoidally, the CTF is analogous to the modulation transfer function. In electron optics the term CTF is often used to refer to the phase contrast transfer function under coherent imaging conditions.

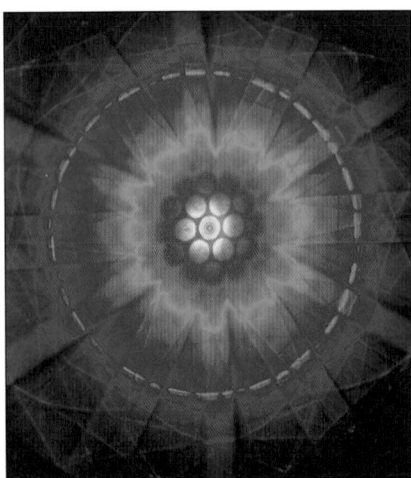

Convergent beam electron diffraction of silicon [111]. Courtesy of Gatan, Inc.

control grid see **gun grid**

conventional transmission electron microscopy

The basic form of operation of a transmission electron microscope in which the core techniques are the formation of bright- and dark-field images and selected-area diffraction patterns from thin specimens using a relatively broad (10-100 μm), approximately parallel, beam of high-energy (60-100 keV) electrons.

convergence angle

1. The angle at which the rays of a converging beam meet. In a TEM the convergence semiangle of the electron beam can be calculated by examining a diffraction pattern of a calibration standard and measuring the diameter of the diffraction discs.
2. The angle between the visual axes of the eyes, typically about 10°.

convergent-beam electron diffraction

Electron diffraction using a converging beam of electrons in a scanning transmission or transmission electron microscope. In convergent beam electron diffraction the angle of convergence of the probe results in a diffraction pattern of disks (in contrast to the spots seen in selected area electron diffraction); as the convergence angle increases, the size of the disks increases and they may then overlap.

convergent-beam electron diffraction pattern

A diffraction pattern produced using convergent beam electron diffraction. A CBED pattern contains detailed contrast governed by the 3D symmetry (unit cell, lattice centering, enantio-morphism, point group and space group) of the specimen, and very accurate thickness, lattice parameter (to 1 part in 10^4), strain and compositional information.

convergent-beam low-energy electron diffraction

Convergent beam electron diffraction using low-energy electrons.

converging lens

A lens that converges parallel light rays and can form a real magnified image.

convolution

The change in one signal by modulation with another. In light and electron microscopy the image of a point in an object is modulated by the point-spread function of the microscope. In scanning probe microscopy, the image of an object may be modified by the shape of the probe tip. Convolution is a mathematical operation, generating a single function $g(y)$ from two others, $f(x)$ and $h(x)$:

$$g(y) = \int h(y - x)\, f(x)\, dx$$

where $f(x)$ and $h(x)$ characterize the optical system and object, respectively, and the arguments y and x may be scalar or vector quantities.

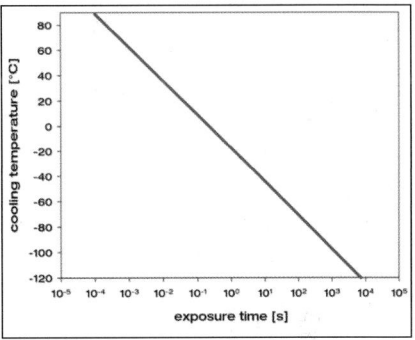

Graph of temperature vs exposure for a dark current of 1e/pixel s in a cooled CCD. Courtesy of PCO AG.

cooled CCD camera

A camera with a charge-coupled device that is cooled below ambient temperature to reduce dark current. The cooling method typically uses air, water, a Peltier device or a cryogen.

cooling stage

A microscope stage used for the investigation of specimens at below ambient temperatures.

copper winding see windings

cornea

The anterior transparent region of the wall of the eye through which light enters the eye. The cornea is the most strongly refracting element in the eye.

correction collar

A rotatable collar that axially displaces lens elements in a light microscope objective. A correction collar allows the spherical aberration correction of the lenses to be adapted in circumstances where the specimen fluctuates in thickness, lacks a coverslip or has the incorrect coverslip thickness, or has an unusual refractive index.

Correction collar on objective. Courtesy of Olympus.

correlation analysis

A mathematical procedure used to compare two signals or images in order to determine their similarity. Correlation analysis may involve autocorrelation or cross-correlation.

corrosion casting

A preparation of a cast that can be revealed by removal of the surrounding material. Corrosion casting is used to examine the vasculature of animals. The blood vessels are perfused with a casting material that polymerizes and the cast is revealed by caustic digestion of all other tissues.

count rate

The number of incident photons or electrons per unit time counted by a detector.

counterpiece

The heel on a glass knife produced when a glass square is fractured across a diagonal into two knives. The height of the counterpieces (typically 0.3 to 1.0 mm) on each knife determines the sharpness and included angle of the blade on the complimentary knife.

coverslip

A thin circular or rectangular sheet of glass, typically placed over a specimen on a slide. Glass coverslips are manufactured in several thicknesses: nominal sizes and thickness ranges are: #1: 0.13-0.17 μm; #1½: 0.16-0.19 μm; #2: 0.17-0.25 μm. The optical design of light microscope objectives takes into account whether or not a coverslip is in the light path: e.g. oil immersion lenses are designed for use with 0.17 μm-thick coverslips, whereas reflected light microscope objectives are designed for use without a coverslip.

critical angle

The angle of incidence above which total internal reflection occurs. Symbol θ_C. The critical angle θ_C is related to the refractive indices n of the media at the interface by the expression: $\sin \theta_C = n_1/n_2$. Critical angle for glass ($n = 1.5$) to air ($n = 1$) is 41.8°; critical angle for TIRF applications using a glass substrate and living cells ($n = 1.38$) is 65°.

critical angle fluorescence microscopy see total internal reflection fluorescence microscopy

critical excitation energy

The minimum energy required to cause ionization of an atom. The critical excitation energy varies with atomic number and the location (orbital shell) of the ejected electron.

critical excitation potential see critical excitation energy
critical Förster distance see Förster radius
critical illumination see source-focused illumination
critical ionization potential see critical excitation energy

critical point

The point at which the density of the liquid phase is equal to that of the vapor phase of a fluid. The fluid is then at its critical temperature, critical pressure and critical volume.

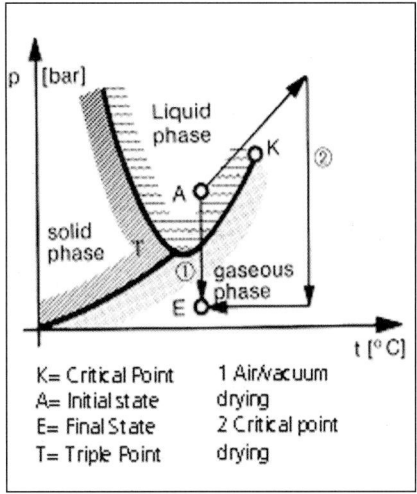

Critical point phase diagram. Courtesy of BAL-TEC AG.

critical-point dryer

A device for specimen processing using the critical point drying method. In a critical point dryer the specimen is placed in a pressure chamber in an organic solvent, e.g. acetone. The specimen is then flushed with a transition fluid, typically CO_2 (critical pressure 7,398,080 Pa, critical temperature 31.1°C) and heated to a temperature several degrees above the critical temperature of the transition fluid; the gaseous transition fluid is then vented.

critical-point drying

The drying of a specimen by heating a surrounding liquid to its critical temperature, thus preventing specimen damage from the effects of surface tension.

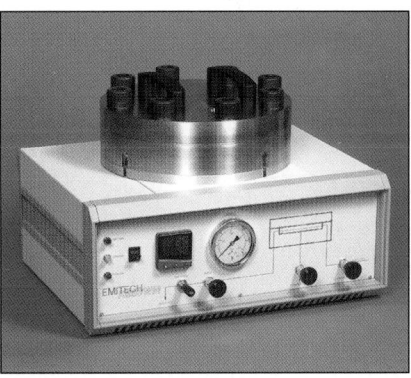

Critical-point dryer. Courtesy of Emitech Ltd.

critical pressure

The pressure at the critical point of a fluid.

critical temperature

The temperature of a fluid at its critical point. A gas cannot be liquified by pressure above its critical temperature.

critical voltage

1. The voltage at which a gas ionizes and forms a plasma.
2. In electron diffraction, the voltage at which it is possible to measure the internal potential of metals.

CrossBeam electron microscope

A brand-specific field-emission scanning electron microscope containing a focused ion-beam system.

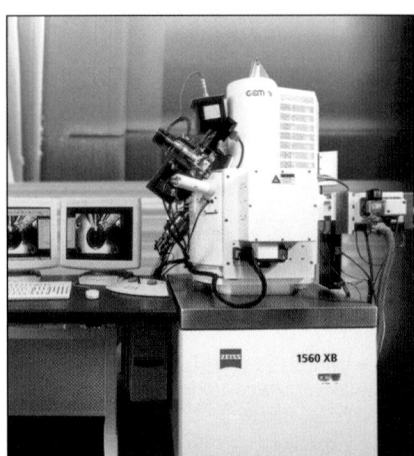

CrossBeam FIB/SEM. Courtesy of Carl Zeiss SMT.

cross-correlation

The correlation of two images or signals. The cross-correlation function C is defined by:
$$C(x^1,y^1) = \iint f^*(x - x^1, y - y^1)G\ (X,Y)\ dx\ dy$$
in which the asterisk denotes the complex conjugate. The Fourier transform of C is equal to F*G, where F and G are the Fourier transforms of f and g respectively.

crossover

The smallest diameter of the beam leaving an electron gun. The crossover is the effective electron source in an electron microscope. Typical crossover diameters are: tungsten filament, 50 μm; lanthanum hexaboride filament, 10 μm; field emitters: 3-15 nm.

cross-section

1. The effective area that an object presents to incident radiation. For example: the area of an atom that produces scattering; the area of an airplane that reflects a radar beam.
2. A section of a specimen cut at right angles to its major axis.
3. A section through a 3D image.

cross-section vise

A vise used in the production of a single cross-section of a stack of wafer specimens.

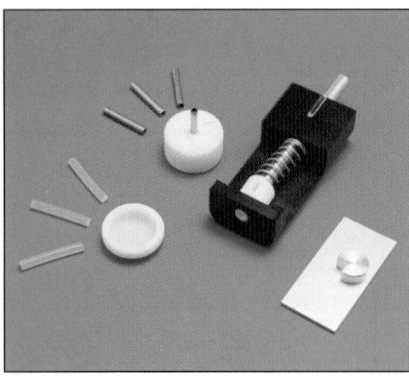

Cross-section vise. Courtesy of E. A. Fischione Instruments, Inc.

cross-sectional transmission electron microscopy

The examination of thin sections of embedded specimens by transmission electron microscopy.

crossed-eyes viewing

A technique for the unaided viewing of parallel stereo pairs of images. In crossed-eyes viewing the eyes are crossed so that the left and right images of the stereo pair are superimposed, producing a stereoscopic image.

crossed polars

The condition in which two polarizers are arranged with orthogonal planes of polarization, giving total extinction of incident light. Only light that changes its state of polarization when passing between the polars, due to interaction with a specimen, is transmitted.

crown glass

A family of optical glasses with low refraction and dispersion, with an Abbe number higher than 55.

cryochamber

A chamber used for low-temperature processing of specimens.

cryoelectron microscope

A scanning or transmission electron microscope with a cold stage for the examination of frozen specimens, typically held at liquid nitrogen or liquid helium temperatures.

cryoelectron microscopy

The use of a cryoelectron microscope.

cryoelectron microscopy, in SEM

The use of a scanning electron microscope to study specimens at low temperatures. CryoSEM is used to examine bulk specimens that cannot or must not be prepared with conventional specimen processing methods, such as fully hydrated biological specimens, low melting point or volatile specimens, and soft solids and liquids, in order to maintain the specimen as close as possible to its natural state. CryoSEM requires a cooling stage, typically at liquid nitrogen or liquid helium

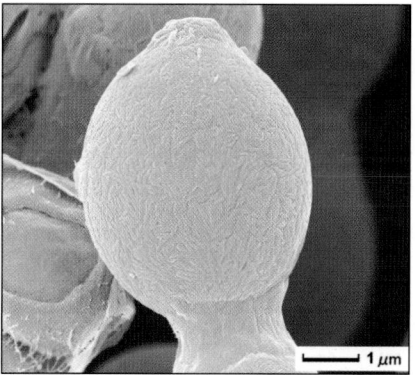

Cryo scanning electron micrograph of blue-cheese yeast. Courtesy of Gatan Inc.

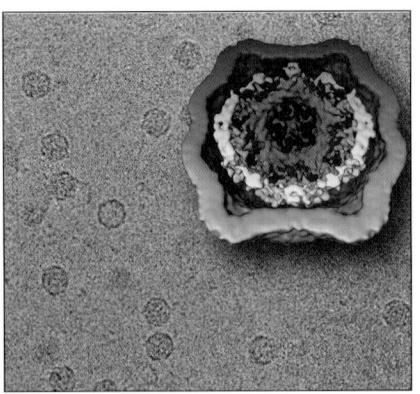

Cryoelectron microscopy. CryoTEM image of Cow Pea mosaic virus in a thin vitrified film, observation under low dose conditions at -180°C. Insert shows 3D reconstruction of the virus particles in the background. Courtesy FEI Applications Laboratory.

Cryoknife in cryoultramictotome. Courtesy of Boeckeler Instruments, Inc.

temperatures. The SEM may be operated at low kV to prevent beam damage.

cryoelectron microscopy, in TEM

The use of a transmission electron microscope to study specimens at low temperatures. CryoTEM requires a cold stage typically at liquid nitrogen or helium temperatures. CryoTEM is typically used for high resolution structural analysis of cryofixed viruses, macromolecules and particles using minimum dose imaging and electron tomography.

cryoelectron tomography

Electron tomography of frozen specimens.

cryofixation

The fixation of water-containing specimens using a cryogen or cooled surface. For optimal cryofixation at normal pressure the specimen must be cooled at a rate greater than $10,000°C\ s^{-1}$ to produce vitrification or minimize ice crystal formation. Standard cryofixation techniques include: immersion or plunge-freezing in a cryogen; metal-mirror cryofixation; propane-jet freezing; and high-pressure freezing.

cryofixation device

Any device used for the cryofixation of specimens, such as a propane jet freezer, high pressure freezer or plunge freezer.

cryofracture see freeze fracture

cryogen

A liquified or solid gas used for cooling specimens, instrument parts or cryofixation. Commonly used cryogens in microscopy are: solid carbon dioxide (dry ice, mp -78.5°C); liquid propane (mp -187.6 °C, bp -42°C,); liquid nitrogen (mp -209.9°C ; bp -195.8°C); liquid helium, (bp -268.9°C).

cryogenic pump

A vacuum pump that traps gases by adsorption to a cryogen-cooled surface.

cryoknife

A knife used for the production of cryosections.

cryonegative staining

The use of negative staining to enhance contrast in cryoelectron microscopy.

cryopreparation and transfer device

A device for the preparation and transfer of specimens into a scanning electron microscope. A cryopreparation device has an evacuated chamber that can be connected to a port at the side of the microscope; the specimen is placed at the end of an arm that moves the specimen through the preparation chamber into the microscope. The chamber contains a liquid nitrogen cooled stage and accessories for specimen preparation, such as fracturing, etching and coating. The device allows transfer from cryogen to microscope without warming or contamination.

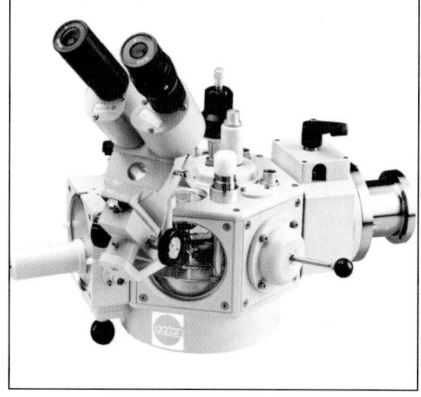

Cryopreparation and transfer device. Courtesy of Gatan, Inc.

cryoprotectant

Any agent that reduces damage to a specimen by ice crystal formation during freezing.

cryosection

A section cut from a frozen specimen.

cryostat

1. A microtome enclosed in a cryochamber used for the production of cryosections for light microscopy and histology.
2. A device for cooling a spectrometer either by conduction, using liquid nitrogen, or electrically, by the Peltier effect.

Cryosections at knife edge of a cryoultramicrotome. Courtesy of Boeckeler Instruments, Inc.

cryosubstitution see freeze substitution

cryotransfer

The transfer of frozen specimens from a cryofixation device to a specimen preparation device and into a microscope at low temperature and at ambient pressure or under vacuum.

cryotransfer holder

A specimen holder for the transfer of frozen specimens from a cryogen or cryoworkstation into a transmission electron microscope. The holder may have a means for maintaining low temperatures, and a shutter to prevent condensation of ice onto the specimen during transfer to the precooled stage of the TEM.

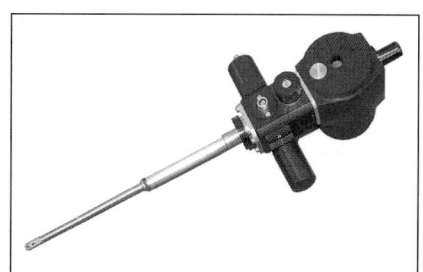

Cryotransfer holder. Courtesy of Gatan, Inc.

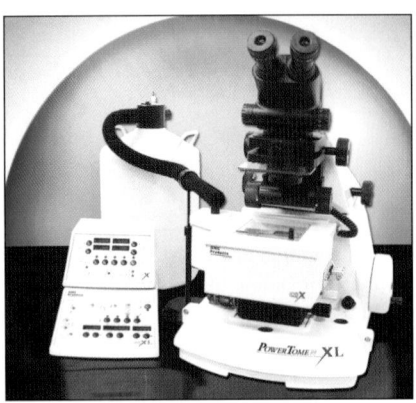

Cryoultramictotome. Courtesy of
Boeckeler Instruments, Inc.

cryoultramicrotome
An ultramicrotome equipped with a cryochamber
for the production of cryosections.

cryoultramicrotomy
The use of a cryoultramicrotome.

crystal scanner see **piezoelectric scanner**
crystal quartz see **quartz**

crystal spectrometer
A spectrometer that uses a crystal to diffract the
input beam and produce a spectrum of
wavelengths, such as a wavelength-dispersive
X-ray spectrometer.

curing see **polymerization**

current centering
An alignment procedure in a transmission electron
microscope that ensures that the objective lens
field is centered around the optical axis. In current
centering, the objective lens current is wobbled
about focus, rotating the image on the screen; the
beam tilt controls are used to center the axis of
rotation.

curvature of field see **field curvature**

cut-off frequency
1. The frequency or wavelength above or below
which light is blocked by an interference filter.
2. The spatial frequency at which the modulation
transfer function approaches zero.

cutting speed
The speed at which the block face passes the knife
edge in a microtome.

cyan fluorescent protein see **fluorescent
proteins**

cylindrical lens
A lens having at least one refracting surface
shaped like a segment of a cylinder, typically used
for magnification in one dimension only.

cytochemistry

The identification and localization of the chemical and molecular constituents of cells by stains, probes and indicators.

cytotoxicity see phototoxicity

D

damping functions

Functions representing the effects of partial temporal and spatial coherence that modulate the contrast transfer function of an optical system.

dark adaptation

The increased sensitivity of the eyes to light after approximately 20 minutes exposure to darkness, resulting from full dilation of the iris and increased amounts of rhodopsin in the retinal rods.

dark charge noise see dark current

dark current

The signal from a detector in the absence of any input from the specimen. Dark current is caused by thermal effects in the detector and its associated electronic components.

dark signal see dark current

darkfield condenser

A condenser that illuminates the specimen with oblique rays of light for darkfield microscopy. Specific designs of oil-immersion darkfield condenser include the paraboloidal and cardioid condensers, but most standard condensers can be used for darkfield microscopy with an annular diaphragm.

darkfield detector see annular darkfield detector

darkfield electron microscopy

The use of a scanning transmission or transmission electron microscope for darkfield microscopy. In standard darkfield TEM the image is formed by using the objective aperture to select electrons scattered at specific angles either by displacing the aperture or by tilting the beam. In a STEM darkfield images are obtained by tilting the beam or by using an annular darkfield detector. Darkfield image contrast is related to specimen

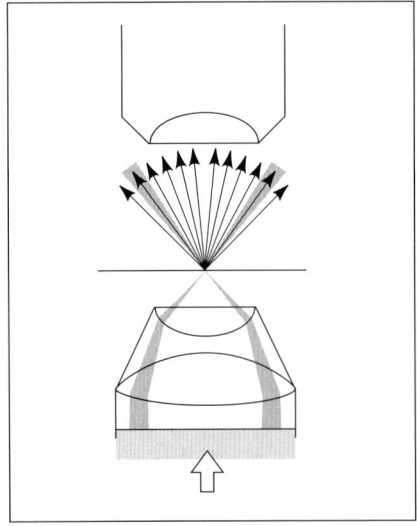

Principle of darkfield condenser in light microscopy. Only diffracted light (arrows) is collected by the objective. Non-diffracted light (gray) misses the lens aperture. From D. Murphy © John Wiley and Sons, Inc.

characteristics that cause specific scattering processes such as crystal orientation (diffraction), thickness and atomic number (mass-thickness).

darkfield light microscope

A light microscope equipped with accessories for darkfield microscopy. The standard accessories are an annular diaphragm in the front focal plane of the condenser, or a special darkfield condenser. The numerical aperture of the condenser should be higher than that of the objective. Light rays from the annular diaphragm are focused onto the pass at an oblique angle. Only light that is diffracted by the specimen is collected by the objective and used to form an image in which objects appear bright against a dark background.

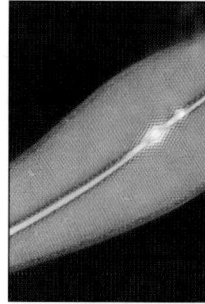

Darkfield light microscope image of diatom (left) and brightfield image (right). Courtesy of Carl Zeiss Ltd.

darkfield microscopy

A form of light or electron microscopy in which only the radiation diffracted by the specimen is used to form the image, therefore the specimen appears bright against a dark background.

darkfield objective

A high numerical aperture light microscope objective with an internal diaphragm to block direct undiffracted light.

darkground microscopy see darkfield microscopy

de Broglie wave

The wave associated with a moving object. The wavelength λ of a de Broglie wave is given by:
$$\lambda = h/p$$
where h is Planck's constant and p is momentum.

de Sénarmont compensator

A type of compensator comprising a fixed quarter-wave plate and a calibrated rotatable analyzer.

dead layer

The p- and n-type regions of a p-i-n semiconductor detector that do not contribute to the charge pulses.

deadtime

The time that a detector and its associated electronic circuitry are unresponsive to signals while recently acquired signals are processed.

deconvolution

A technique to remove unwanted or background information from an image, signal or spectrum. Deconvolution is a mathematical operation by which an image or other signal can be retrieved from a degraded signal when the latter is the convolution of the original signal and another function.

dedicated microscope

Any microscope that is designed and used solely for one technique, such as a Raman microscope or microprobe.

deflection coils

A pair of coiled-wire electromagnets used to deflect an electron beam. In electron microscopes, deflection coils are usually arranged as a double set, one above the other, in horizontal arcs around the column with one member of each pair attracting and the other repelling the beam. When the upper set of coils bends or tilts the beam by an angle α, the lower set can be used to tilt the beam by α in the opposite direction, producing a pure sideways shift of the beam, parallel to its original direction.

deflection mode see constant height mode

defocus spread

The spreading of the value of defocus that occurs with a polychromatic electron beam.

degassing

The removal of gases by vacuum and/or heat treatment.

dehydration

The removal of water. Most biological specimens and some materials are dehydrated before coating or embedding.

depth discrimination see depth of focus, depth of field

depth of field

The distance along the optical axis which an object in the specimen can be moved while remaining in acceptable focus in the image.

depth of field, in TEM

The depth of field of a transmission electron microscope is typically greater than the thickness of most ultrathin sections and foils because of the very small objective apertures used, so the whole specimen is in focus in the image. As SEMs and STEMs have no post-specimen lenses the terms depth of field and depth of focus are synonymous. The depth of field of a SEM at low magnifications is typically several millimeters. In an SEM the electron beam is a focused spot moving across the surface of the specimen; any adjacent region above or below this point will also be 'in focus' providing the diameter of the converging or diverging beam at that point does not exceed twice the size of the pixels in the image.

depth of field, in LM

The depth of field Z can be calculated from the expression:

$$Z = n\lambda/NA^2$$

where n is the refractive index of medium between lens and object, λ is wavelength of light in air, and NA is numerical aperture of objective lens. The depth of field can be measured by imaging a periodic specimen or a micrometer scale placed at an oblique angle to the object plane. For a low magnification 0.1 NA objective the DOF is about 55 μm; for a high magnification 1.4 NA objective the DOF is 0.38 μm.

depth of focus

The distance along the optical axis either side of the image plane within which objects are in acceptable focus when the position of the object is unchanged.

depth of focus, in LM

The depth of focus Δz of a light microscope can be estimated from the expression:

$$\Delta z = \lambda/4n(1-[1-(NA/n)^2]^{1/2})$$

where λ is wavelength, NA is numerical aperture and n is refractive index.

depth of focus, in SEM see **depth of field, in SEM**

depth of focus, in TEM
The depth of focus of a transmission electron microscope is typically on the order of several meters, allowing the user to focus the image on the viewing screen and still get a perfectly focused image recorded on the film or detector many centimeters below the screen.

derivatized tips see **chemical tips**

descanned mode
An optical configuration in confocal microscopy in which the light emitted by the specimen passes to the detector via the scanning head beamsplitters.

desiccant
A hygroscopic material.

desiccator
A device that keeps water vapor from specimens.

destructive interference
Interference in which the resultant wave has an amplitude lower than that of the interacting waves.

detection limit
The smallest object that can be imaged by a microscope or detector. The detection limit should not be confused with image resolution: a light microscope in combination with a sensitive CCD camera can detect single photons, but may not be able to resolve two closely spaced photons.

detection/detective quantum efficiency see **detector quantum efficiency**

detector
A device that detects radiation or charged particles and is used to form an image or a spectrum.

detector collimator
A collimator at the front of an energy-dispersive X-ray spectrometer. The collimator has a slit or circular aperture and is made of a high-Z material coated with a low-Z material to reduce X-ray

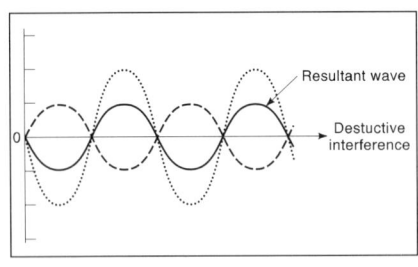

Destructive interference of waves. From D. Murphy © John Wiley and Sons, Inc.

fluorescence and may have baffles to reduce entry
of backscattered electrons.

detector icing

The condensation of water vapor on a cooled
detector. Detector icing is a potential problem with
all detectors that operate below 0°C. To reduce
detector icing, cooled CCD cameras and X-ray
spectrometers usually operate in a vacuum behind
a hermetically sealed transparent window.

detector pinhole see pinhole aperture

detector quantum efficiency

The efficiency with which a detector converts
incoming to outgoing signals. A perfect detector
has a QE of 1, meaning that each incident photon
is converted into an outgoing electrical pulse.

detector shutter see shutter

detector window

A thin vacuum-tight transparent film at the front of
an X-ray spectrometer. Detector windows are made
of a variety of materials including beryllium,
silicon nitride, diamond or polymer.

developer

A chemical solution used to produce an image in
exposed film negatives or photographic papers. A
developer reduces the latent image of exposed
silver bromide to black metallic silver. A developer
typically contains: a developing agent, e.g.
hydroquinone; an alkali (accelerator), e.g. sodium
hydroxide; a preservative, e.g. sodium sulphite; a
restrainer, e.g. potassium bromide, and other
components such as a wetting agent and anti-
sludging agent.

developing tank

1. A small light-proof container for the development of 35-mm film.
2. An apparatus for TEM film sheet development,
equipped with temperature control, nitrogen burst
agitation, timers, and distilled water supply,
situated in a darkroom.

deviation parameter see excitation error

Diamond knife. Courtesy of Boeckeler Instruments, Inc.

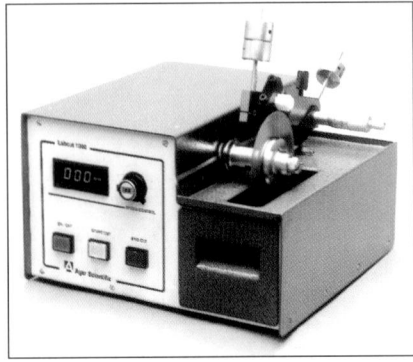

Diamond saw. Courtesy of Agar Scientific.

Dewar flask

A vessel for the storage of cryogenic liquids with walls made of two thin layers of glass or metal separated by a vacuum.

diamond knife

A knife made from diamond used to cut thin sections in an ultramicrotome. Diamond knives have sharp, durable blades and are capable of producing extremely thin sections (>30 nm) of soft or hard biological and materials specimens.

diamond saw

A saw with diamond teeth used for the precision cutting of hard materials.

diaphragm

A mechanical device that limits the cross-sectional area of a beam of radiation. Diaphragms in optical devices are usually circular and often adjustable, as in the iris diaphragms of a camera and condenser.

diaphragm pump

A mechanical, oil-free pump that pumps gases by the flexing of a diaphragm.

diascopic mode

The use of transmitted illumination to form an image in a light microscope.

dichroic crystal

A crystal that shows two different colors of transmitted light when viewed at different angles due to dichroism. Examples: tourmaline, herapathite (quinine sulfate periodide, the original optically active component of Polaroid film).

dichroic filter

A general term for an optical filter that transmits one range of wavelengths of light or color and absorbs or reflects other wavelengths.

dichroic mirror

A filter that reflects light of one range of wavelengths and transmits another. A dichroic mirror is used in a fluorescence filter cube to reflect excitation wavelengths onto the specimen and transmit emission wavelengths to the eyepieces.

dichroic polarizer

A polarizer made of dichroic material such as Polaroid.

dichroism

1. The selective absorption of one of the two orthogonal polarization states of light by optically anisotropic materials. In dichroic crystals light waves with a polarization state perpendicular to the optic axis are absorbed whereas parallel polarization states are transmitted; in a wire-grid polarizer and in Polaroid sheets, the parallel states are absorbed.
2. The selective absorption of specific wavelengths of light by optically anisotropic materials. Some crystals, such as tourmaline, show two different colors when illuminated along different axes by white light.

dielectric

A non-conducting material in which an applied electric field can cause a displacement of electrons but not a flow of charge. Dielectrics do not absorb light and are commonly used in interference filters.

difference image

An image produced by subtracting one image from another.

differential cross-section

The change of electron scattering cross-section with solid angle of scatter: $d\sigma/d\Omega$. The differential cross-section embodies all the factors that control electron scattering (and thus image contrast) in an electron microscope. The differential cross-section is strongly proportional to the electron charge and the atomic number and inversely proportional to the beam energy and the angle of scattering angle.

differential interference contrast

The generation of image contrast in double-beam interference microscopy by using beams separated laterally by a distance below the resolving power of the microscope.

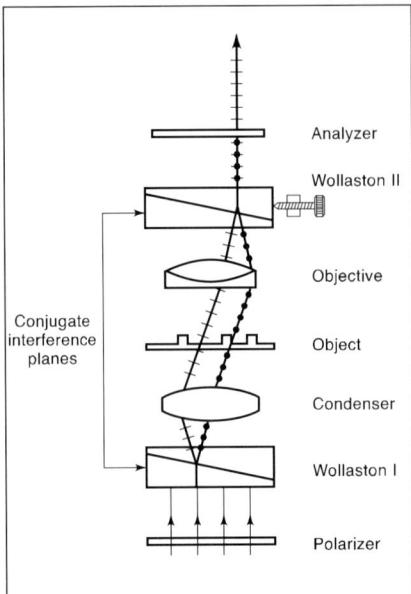

Optical components and pathways in a DIC light microscope. From D. Murphy © John Wiley and Sons, Inc.

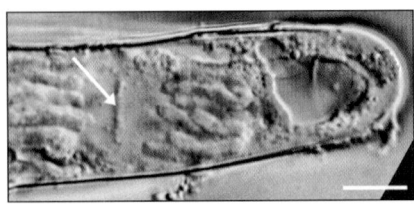

DIC image of dividing Spiderwort stamen hair cell showing cell plate (arrow). Scale bar = 10 μm. *Microscopy and Analysis* 2001.

differential interference contrast light microscope

A type of light microscope with accessories for differential interference contrast. The accessories required for DIC are a sub-stage linear polarizer, a monochromatic filter (e.g. green), a Nomarski or Wollaston prism in the front focal plane of the condenser, a Nomarski or Wollaston prism in the back focal plane of the objective, and an analyzer in the tube. The polarizer and analyzer are orientated at right angles, typically east-west and north-south respectively, to produce maximum extinction. Polarized monochromatic light is split into pairs of closely spaced, parallel, orthogonally polarized E and O rays by the condenser prism. The shear distance of the rays is below the resolving power of the objective. As the ray pairs pass through the specimen they may undergo a relative phase shift due to gradients of specimen refractive index. After collection by the objective, the ray pairs are recombined by the objective prism generating linearly polarized rays (if they have no phase shift) or elliptically polarized rays (if there is a relative phase shift); the former are then blocked by the analyzer, whereas the latter pass through to the intermediate image plane where they interfere with other transmitted rays producing an amplitude contrast image with bright phase-retarding objects on a dark background. The objective prism can add a bias retardation to the ray pairs, increasing background intensity to gray with corresponding intensity increases of phase gradients, resulting in a pseudo-relief image. If a full wave plate is placed in front of the analyzer, an image with interference colors is produced when using white light.

differential interference contrast light microscopy

The use of a differential interference contrast microscope

differential pumping aperture see pressure-limiting aperture

diffracted wave

A wave that is diffracted by a specimen or a phase object.

diffracting crystal

A crystal that diffracts radiation in a spectrometer. Diffracting crystals used in wavelength-dispersive spectrometry are made from crystalline materials with different plane spacings to diffract a specific range of wavelengths covering all elements from boron to uranium. Examples of WDX crystals and their analyzing ranges are: thallium acid phthalate: 500 eV -1.6 keV; pentaerythritol 1.5 - 5 keV; lithium fluoride: 4-15 keV.

diffraction

The scattering of waves by an object. Although diffraction is the basis of image formation of most specimens in light and electron microscopes, the term is commonly used to describe the interaction of waves with a periodic or crystalline object that produces a diffraction pattern. In a transmission electron microscope, diffraction from crystalline specimens results in coherent scattering in specific directions related to the crystal-plane spacing, while non-crystalline specimens give rise to more diffuse diffraction patterns.

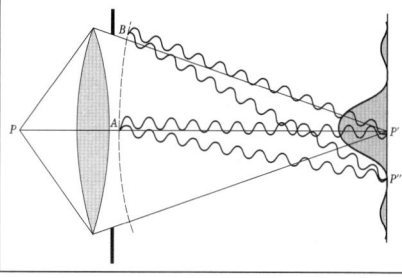

Principle of formation of a diffraction pattern of a point source P by a lens. On-axis and off-axis waves interfere constructively (P') or destructively (P'') in the image plane, forming an Airy pattern. From D. Murphy © John Wiley and Sons, Inc.

diffraction aperture

1. In optics, an aperture that diffracts a beam, e.g. the aperture of a lens or a slit diaphragm.
2. Synonym for selected area diffraction aperture.

diffraction centering

An alignment procedure for a transmission electron microscope in which the direct beam in a diffraction pattern is centered on the optical axis using the projector lens.

diffraction contrast

Contrast that derives from the use or exclusion of diffracted beams in image formation.

diffraction coupling see diffraction mode

diffraction disk

1. A synonym for Airy disk.
2. A disk produced by convergent beam electron diffraction.

diffraction grating

A transparent or reflective material ruled with

equally spaced slits or grooves that diffract and disperse light. The distance between orders of bands in the diffraction pattern is related to grating spacing d and the wavelength λ of normally incident light by the expression: $d = n\lambda/\sin\theta$, where n is diffraction order (1, 2, 3, etc) and θ is the angle subtended by orders n and zero at the grating.

diffraction lens, in TEM

One of the intermediate lenses in a transmission electron microscope that takes as its object the diffraction pattern in the back focal plane of the objective lens and projects the pattern onto the viewing screen. The diffraction lens may also be used for low magnification imaging.

diffraction limit

The limit to the resolving power of an imaging system and the resolution of its images set by the effects of diffraction.

diffraction mode

1. A synonym for conoscopic mode.
2. The use of a diffraction pattern as the object for an electron energy-loss spectrometer.

diffraction pattern

A pattern of spots, disks, rings or lines caused by the diffraction of incident radiation by periodic structures in a specimen. A diffraction pattern is formed in the back focal plane of a microscope objective.

diffraction spot

A spot or region of high intensity in a diffraction pattern.

diffractogram

An image of a diffraction pattern.

diffractometer

1. A device that produces an optical diffraction pattern from an image recorded on film. The diffraction pattern is generated by passing a laser beam through an area of interest in a negative.
2. An instrument that produces an X-ray diffraction pattern.

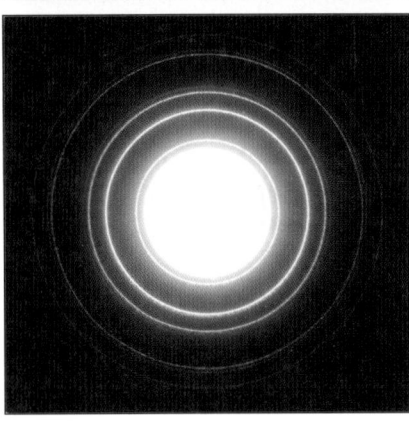

Diffraction patterns. Top: electron diffraction of whole crystal. Courtesy of Carl Zeiss SMT. Bottom: electron diffraction of powdered crystal. Courtesy of Agar Scientific.

diffuse reflectance FTIR spectroscopy

A mode of Fourier-transform infrared spectroscopy used in the analysis of powdered samples. The beam is directed onto the specimen using a series of planar and elliptical mirrors that also collect the diffusely reflected beam and direct it into the spectrometer.

diffuse reflection

The reflection of light from an irregular surface, causing a change in the spatial distribution of the constituent rays.

diffuser

A glass plate that randomly scatters light. A diffuser may be placed in front of a light source to improve the uniformity of illumination.

diffusion pump

A vacuum pump that uses pressurized streams of oil vapor to trap gas. The base of a diffusion pump contains a reservoir of oil that is boiled to release a vapor that is then forced out of a vertical series of jets onto the water-cooled sides of the pump where the oil condenses and returns to the reservoir. During this process the oil vapor traps gas molecules which are removed by a backing pump. Diffusion pumps can produce a vacuum below 10^{-4} Pa.

digital camera

A camera that produces a digital image.

Digital light microscope. Courtesy of Nikon.

digital diffractogram

A diffraction pattern produced by Fourier transformation of a digital image.

digital image

An image in which each point or pixel has a discrete value.

digital microscope

A term used variously to describe a light microscope that is operated via a computer and monitor, and/or uses digital imaging technology.

digital pulse processor

A solid-state pulse processor that combines the

functions of the analog pulse processor and analog-to-digital converter in X-ray spectrometers.

digitization

The process of changing an analog image or signal with continuously variable values into one with discrete values.

dimpler see **dimpling grinder**

dimpling

The formation of a concave depression. Dimpling, typically by wet abrasion with fine grits, is a technique used to produce a materials specimen with a central thinned region for microscopy and an outer thicker rim for handling.

dimpling grinder

A device used for dimpling of materials specimens. A dimpler contains a vertical grinding wheel that abrades the specimen as it is rotated and oscillated horizontally to produce a central thin area.

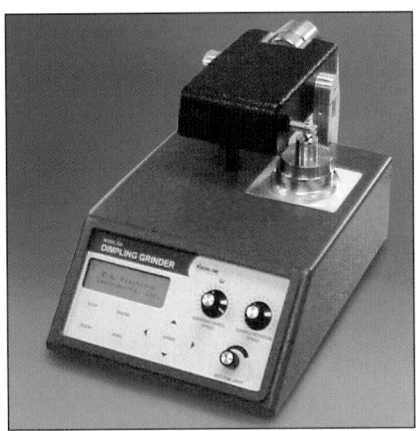

Dimpling grinder. Courtesy of E. A. Fischione Instruments, Inc.

DIN film rating

The Deutsches Institut für Normung standard for film speed, now superseded by the ISO rating.

diode gun

An electron gun with two elements, an anode and cathode, commonly used in coating units and some video tubes.

diode laser

A light-emitting diode with a feedback cavity to produce stimulated emission of light. The emitted wavelength is specific to the composition and dimensions of the light-emitting diode.

diopter

A measure of the optical power of a lens, equal to the reciprocal of its focal length in meters.

diopter adjuster

A mechanism in a light microscope eyepiece for adjusting the position of the eyelens to compensate for differences in dioptric power between the eyes of the observer(s).

dioptric

Regarding an optical system containing refractive elements.

dip-pen nanolithography

The use of an atomic force microscope to deliver molecules to a substrate, forming nanometer-sized patterns. In DPN, transfer relies on the formation of a droplet of aqueous or organic solvent between the substrate and a probe coated with the patterning material; as the probe moves in contact mode over the substrate dissolved molecules attach to the substrate by chemisorption or electrostatic interaction.

direct beam

A beam that is not diffracted by a specimen.

direct labeling

The attachment of a cytochemical probe to a specimen without the use of an intermediate molecule or bridge.

disc microtome

A design of microtome, used for histology, in which the block sits on a disc that rotates past the knife edge.

discrete dynode electron multiplier

A photomultiplier with separate dynodes.

discrete Fourier transform

A form of Fourier transform employed in image analysis when the dataset is in digital form, having discrete values.

dispersion

The change of refractive index with wavelength or energy. Dispersion results in the separation of a beam of polychromatic radiation into its monochromatic components, leading to chromatic aberration. For most transparent materials, refractive index decreases with wavelength: a crown glass prism disperses white light into colors, bending red less than blue because its refractive indices for these colors are 1.53 and 1.55, respectively.

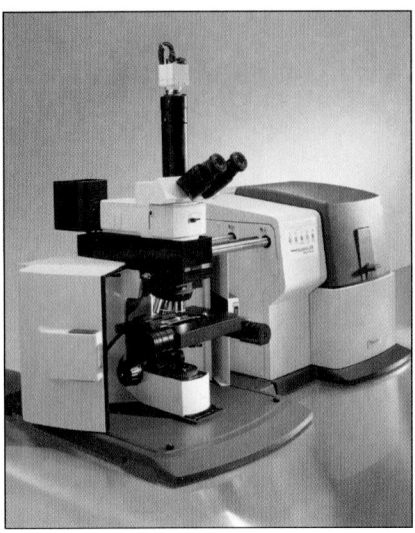

Dispersive Raman microscope. Courtesy of Thermo Electron Corporation.

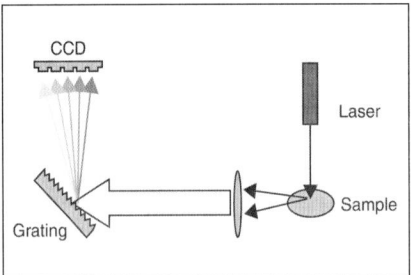

Principle of dispersive Raman microscopy. Courtesy of Thermo Electron Corporation.

dispersive infrared microscopy

Infrared microscopy in which the emitted radiation is dispersed and analyzed using an optical grating spectrometer.

dispersive Raman microscopy

The use of a dispersive Raman microscope.

displacement damage see knock-on damage

dissection microscope

A stereomicroscope used for the dissection of specimens.

dissipative absorption

Absorption in which there is no accompanying emission of radiation.

distortion

An aberration of lenses caused by a change in lateral magnification with distance from the optical axis. The two common types of distortion are barrel and pincushion. Distortion depends on the third power of the off-axis distance and is independent of the angle at the object.

diverging lens

A lens that diverges parallel light rays and can form a virtual demagnified image.

doping

The addition of atoms, the dopant, to a semiconductor detector by electrical diffusion or ion implantation.

dot mapping

The generation of an image of the distribution of X-ray intensity as a function of the position of the scanning electron beam. Dot maps provide qualitative information about the distribution of X-rays from a single element. Qualitative dot mapping has generally been superseded by quantitative mapping and spectrum imaging techniques.

double-beam interference

Interference of two beams of radiation, typically produced from the same source.

double-beam interference microscopy

The use of separate object and reference beams in interference microscopy.

double deflection coils see **deflection coils**

double diffraction

The rediffraction of a diffracted beam. Double diffraction may occur in the same crystal or when a diffracted beam enters an adjacent crystal.

double focusing

A property of a magnetic-sector prism or a sequence of quadrupole lenses that causes electrons in mutually orthogonal planes to be brought to the same focus.

double refraction

A property of optically anisotropic materials causing light to be refracted into two orthogonally plane-polarized rays, the ordinary ray and the extraordinary ray.

double-tilt holder

A specimen holder that allows tilting of the specimen on two mutually orthogonal axes.

drift tube

An evacuated tube for the passage of electrons or ions in spectrometers.

drifting see **doping**

dry-mass interferometry

The use of a transmitted-light interference microscope to calculate the dry mass of thin biological specimens (e.g. cells). Since the specific refractive increments of most biological macromolecules are similar, the measurement of cellular refractive index from optical path differences yields the dry mass.

Double refraction in a calcite crystal. From D. Murphy © John Wiley and Sons, Inc.

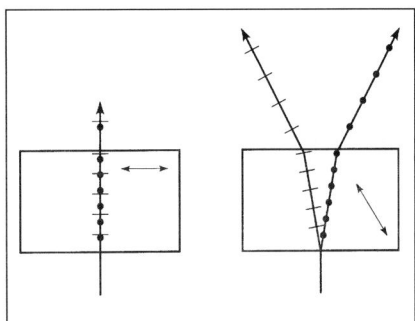

Principle of double refraction in calcite. Double headed arrow is optic axis. Left: A ray perpendicular to optic axis emerges as parallel E and O rays. Right: At oblique incidence, E and O rays diverge. From D. Murphy © John Wiley and Sons, Inc.

dry objective

A light microscope objective designed for use without an immersion medium. Maximum numerical aperture is ~0.95.

dry rotary pump

A type of rotary pump that operates without oil, such as a scroll pump.

dual-beam electron microscope

A type of electron microscope that has a focused ion-beam to cut or mill a specimen and an electron beam to form secondary electron or scanning transmission electron images of the processed specimen. The ion beam is produced by a gallium liquid-ion metal source.

dual-beam interference see double-beam interference
dual-wavelength probe see ratiometric probe
D wave see diffracted wave

Dual-beam electron microscope (FIB-SEM). Courtesy of FEI Company.

dwell time

The time that a probe spends at a given point on the specimen during an imaging or analytical procedure.

dye

A molecule that binds to and gives color to a specimen.

dynamic force microscopy

The use of an atomic force microscope with an oscillating cantilever and probe. Changes in the dynamic properties of the cantilever (e.g. frequency, amplitude and phase) due to tip-sample interactions are measured.

dynamic mode

A mode in scanning probe microscopy in which the the cantilever or probe is vibrated and image contrast is based on changes in the resonant frequency or phase of the cantilever oscillation.

dynamic range

The ratio of the maximum to minimum values in an image or signal.

dynamical diffraction

Electron diffraction in which the beam is rediffracted several times, typically by thicker specimens.

dynode

A positively charged electrode that attracts electrons, generating secondary electrons.

E

edge filter

1. An optical filter that transmits wavelengths of light above (longpass) or below (shortpass) a certain specification.
2. An image filter that enhances the edges of objects.

edge see absorption edge, ionization edge

einzel lens

A three-electrode electrostatic lens in which the outer electrodes are held at the same potential as the anode of the gun. An einzel lens has only one free parameter, the potential of the central electrode, which may be higher or lower than that of the outer electrodes.

elastic cross-section see elastic scattering cross-section

elastic scattering

The scattering of radiation with no or negligible loss of energy.

elastic scattering cross-section

The cross-section presented by an atom that yields an elastic scattering event. The elastic cross-section increases with atomic number Z.

electric field vector

A vector that describes the direction and amplitude of vibration of the electric field along the axis of an electromagnetic wave.

electric force microscopy

A type of scanning probe microscopy that measures and maps the gradient of the electrical field between the probe and the specimen. In electric force microscopy a conductive tip is used to detect electric fields. Typically a two-pass method is used: first the topography of the specimen is imaged; the tip is then moved away from the specimen and a second scan, tracing specimen topography, is used to measure electric fields.

electrical noise

Noise that is produced by the electronic circuitry of a detector.

electrolytic thinning

The thinning of a metal specimen by the electrochemical action of electrolyte. Electrolytic thinning produces a smooth surface as higher current densities are associated with raised features leading to more rapid release of material.

electroluminescence see cathodoluminescence
electrolytic thinning see electropolishing
electromagnetic lens see magnetic lens

electromagnetic radiation

Radiant energy produced by non-uniformly moving or accelerated charges. Electromagnetic radiation travels at the speed of light in the form of electromagnetic waves with oscillating electric and magnetic fields. The properties of the radiation are dependent on its wavelength or energy.

electromagnetic spectrum

The spectrum of electromagnetic radiation, extending from the longest wavelengths (radio waves), through the visible region (light), to the shortest wavelengths (gamma rays).

electromagnetic wave

The transverse wave associated with electromagnetic radiation. An electromagnetic wave is made up of electric and magnetic fields that oscillate at right angles to each other and to the direction of propagation.

electron

A negatively charged elementary particle. Symbol e. The charge carried by an electron is $1.602,177 \times 10^{-19}$ coulomb. The rest mass m_0 of an electron is 9.109389×10^{-31} kg. The wavelength λ of an electron is related to the accelerating voltage V in

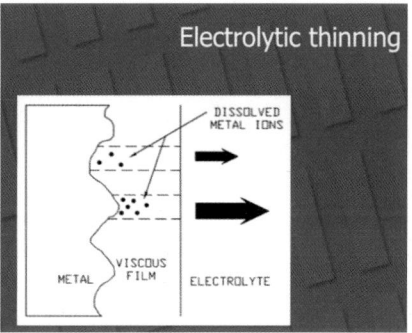

Electrolytic thinning

Principle of electrolytic thinning.
Courtesy of E.A. Fischione Instruments, Inc.

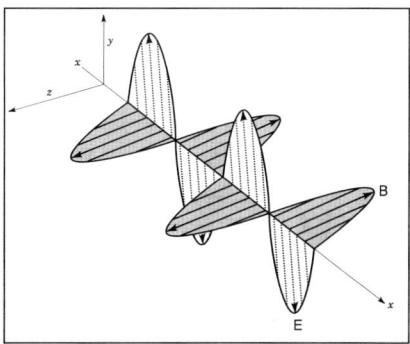

Schematic of electromagnetic wave showing the orthogonal electric and magnetic vectors. From D. Murphy © John Wiley and Sons, Inc.

an electron microscope by the expression:
$$\lambda = 1.226/V^{-1/2} \text{ nm.}$$
This formula does not take account of relativistic effects and is a useful approximation only for scanning and low-voltage electron microscopes operating at up to about 20 kV: at 1 kV, $\lambda = 38.8$ pm; at 20 kV, $\lambda = 8.67$ pm. At the higher accelerating voltages used in transmission electron microscopes the mass of an electron increases due to relativistic effects, and hence the wavelength decreases. The relativistic wavelength λ_r is given by the expression:
$$\lambda_r = 1.226/[V + 0.978 \times 10^{-6} V^2] \text{ nm.}$$
At 60 kV, $\lambda_r = 4.87$ pm; at 100 kV, $\lambda_r = 3.7$ pm; at 400 kV, $\lambda_r = 1.64$ pm.

electron backscatter diffraction

Kikuchi diffraction using backscattered electrons. Electron backscatter diffraction is performed in a scanning electron microscope by tilting a polished bulk specimen at a high angle (~70°) with respect to a stationary electron beam at 20-40 kV. Electrons backscattered from the superficial layers of the specimen are diffracted by local crystal planes and form an electron backscatter pattern on a phosphor screen which is recorded by a CCD camera. EBSD reveals information on crystal orientation, grain boundaries and phase distribution.

electron backscatter diffraction pattern
see **electron backscatter pattern**
electron backscatter Kikuchi diffraction
see **electron backscatter diffraction**

Electron backscattered diffraction pattern of germanium crystal. Courtesy of Oxford Instruments.

electron backscatter pattern
The diffraction pattern produced by electron backscatter diffraction.

electron backscattered diffraction see electron backscatter diffraction

electron beam evaporator
A device for the evaporation of material by bombardment with an electron beam. A typical electron beam evaporator, used in coating units and freeze-fracture devices, contains a tungsten cathode that emits electrons that are accelerated

and focused towards an anode containing the target material to be evaporated.

electron beam-induced conductivity /contrast see **electron beam-induced current**

electron beam-induced current
A flow of current through a specimen induced by an incident electron beam. Electron beam-induced current is an SEM technique used for the detection of defects in materials such as semiconductors. The specimen is placed on an insulator and an electrode is attached to the specimen; as the beam scans the specimen, currents are generated across semiconductor junctions. The induced current is used to modulate the SEM display producing an image as the beam scans the specimen. Defects produce dark contrast; normal areas appear bright.

electron biprism
An electron beamsplitter comprising a thin (~0.5 μm) positively charged (10-100 V) metal-coated glass wire lying between two metal plates. In use, the wire is positioned orthogonal to the electron microscope optical axis, deflecting electrons passing either side towards the optical axis and creating two virtual sources of electrons.

electron bombarded charge-coupled device
A type of CCD with an image intensifier in front of the pixel array. Photons incident of the intensifier produce photoelectrons that are accelerated by a high voltage supply down onto the CCD, increasing its sensitivity.

electron channeling
The facilitated movement of electrons through columns of atoms which act as a lens so reducing electron scattering.

electron Compton profile see **Compton profile**

electron Compton scattering see **Compton scattering**

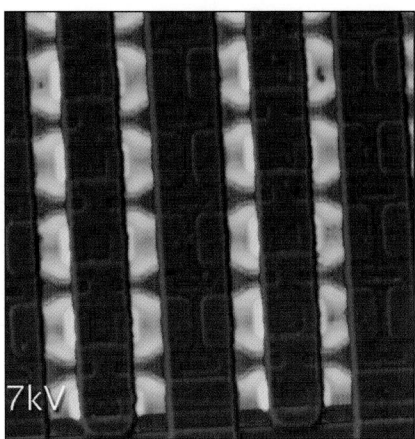

Electron beam induced current. Overlaid EBIC (blue) and SE (red) images from semiconductor amplifier reveals details of the junction beneath the metallization. Courtesy of Gatan, Inc.

electron crystallography

The study of the structure of crystalline materials and macromolecules by electron diffraction.

electron diffraction

The formation of a diffraction pattern using incident electrons. In electron microscopy, diffraction is the primary technique for the analysis of the microstructure of materials.

electron diffraction pattern

An image produced by the diffraction of electrons by a specimen in an electron microscope.

electron dose

The number of incident electrons per unit area.

electron emitter

A common term for the filament of an electron gun.

electron energy-loss spectrometer

An electron spectrometer that analyzes the energy distribution of electrons inelastically scattered by the specimen. Electron energy-loss spectrometers use electrostatic and/or magnetic prisms in a variety of configurations; the two most widely used types are the omega filter and the Gatan imaging filter.

electron energy-loss spectroscopy

The identification of atoms and their electronic structure by analysis of the energy losses of scattered incident electrons. Electron energy-loss spectroscopy is usually performed in a TEM or STEM equipped with an in-column (omega filter) or a post-column (Gatan imaging filter) electron spectrometer that generates an electron energy-loss spectrum. The energy resolution of EELS is <0.2 eV; the spatial resolution is determined by the width and chromatic aberration of the electron probe.

Rectangular CCD array (arrow) used as EELS detector. A 2D spectrum readout from the detector and the EELS spectrum are shown in the background. Courtesy of Gatan, Inc.

electron energy-loss spectroscopy detector

A detector placed after an EEL spectrometer or filter designed to capture electrons. EELS detectors have included film, SIT cameras, and scintillators coupled to photomultiplier tubes (serial detection) or photodiode arrays (parallel detection). Most current detectors comprise a scintillator coupled to a CCD array for optimum linearity and dynamic range.

electron energy-loss spectrum

A spectrum produced by electron energy-loss spectroscopy. The characteristic features of an electron energy-loss spectrum are: the zero-loss peak; the low-loss region; the high-loss region; and the electron Compton profile. A peak in an electron energy-loss spectrum represents an ionization event that has occurred in a specific inner atomic shell (K, L, M, N, O). Since the binding energy of atomic shells is known, ionization edges identify specific elements. The high-loss region contains most of the peaks that are specific to and identify the elements in the specimen; the height of a peak is proportional to the number of atoms that give rise to that peak.

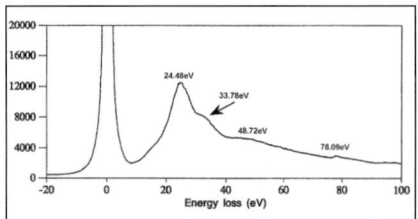

EELS spectrum of aluminum oxide on Ni Cr alloy. *Microscopy and Analysis* 2005.

electron flight simulator

A software program that simulates the trajectories of electrons through a specimen.

electron gun

An electrical device that generates a beam of electrons. Electron guns are used in electron microscopes, in coating units and in video tubes.

electron gun, of electron microscope

The electron gun used to generate the electron beam in all electron microscopes. The two most commonly used designs are thermionic guns and field-emission guns.

electron-hole pairs

Charge carriers consisting of an electron (the negative charge) and its complimentary vacancy or hole (the positive charge).

electron holography

Holography using beams of electrons. The standard technique is off-axis electron holography.

electron lens

A lens containing a region in which a magnetic or

electrostatic field is confined. In round magnetic lenses, the field is rotationally symmetric around a straight optic axis; in a quadrupole lens the field has planes of symmetry and antisymmetry.

electron microprobe

A scanning electron microscope with a configuration and electron optics specifically designed for quantitative X-ray microanalysis. An electron microprobe typically has detectors for secondary and backscattered electron imaging, and energy- and wavelength-dispersive X-ray spectroscopy.

electron microscope

A microscope that uses a beam of electrons to form an image or investigate a specimen. Electron microscopes operate under vacuum as electrons would be scattered by gases in the column. The key features of an electron microscope are: an electron gun; magnetic lenses to focus the beam and magnify images; detectors for imaging and spectroscopy. See: scanning electron microscope; scanning transmission electron microscope; transmission electron microscope.

electron microscope grid see support grid
electron microscope microanalyzer see electron microprobe

electron microscopy

The use of an electron microscope.

electron monochromator

A monochromator placed below the electron gun of an electron microscope, typically of the Wien filter type.

electron-multiplying charge-coupled device

A low light-level CCD with a gain register, placed after the serial register, in which an electric field causes electron multiplication by impact ionization of the silicon oxide.

electron nanodiffraction

Electron diffraction of nanometer-sized areas of crystalline specimens.

Electron multiplying CCD camera. Courtesy of Hamamatsu.

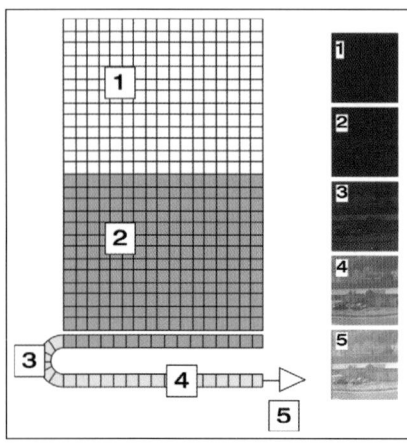

Principle of electron multiplying CCD: sequence on chip (left) and effect on image (right). 1. Imaging area. 2. Storage registers. 3. Transfer to gain registers. 4. Signal amplification. 5. Final image. Courtesy of PCO AG.

electron optics

The study of the properties of electrons and electron imaging systems.

electron-probe (X-ray) microanalysis see microanalysis
electron-probe microanalyzer see electron microprobe

electron Ronchigram

A Ronchigram produced in an electron microscope. The Ronchigram is the disk of undiffracted electrons at the center of a convergent-beam electron diffraction pattern of a specimen in a TEM/STEM. The Ronchigram expresses all the probe-forming optics simultaneously and is the preferred method for examining electron-probe alignment, spherical aberration, astigmatism, overfocus and underfocus. When the electron probe is perfectly aligned and focused, the disc is of uniform intensity.

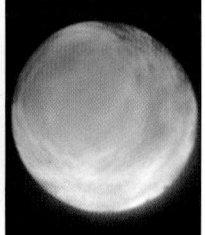

Electron Ronchigrams of gold islands on carbon. Left: severe astigmatism in condenser at focus; right: in focus. *Microscopy and Analysis* 2002.

electron scattering see scattering

electron source

1. A synonym for the filament of an electron gun.
2. The real or virtual origin of an electron beam in an electron microscope.

electron spectrometer

A spectrometer for the analysis of electrons.

electron spectroscopic imaging

The use of electrons with a specific energy bandwidth to form an image in an energy-filtering transmission electron microscope. Electron spectroscopic imaging is used to feature the distribution of specific elements in a specimen, or improve the resolution of images of thick specimens degraded by chromatic aberration.

electron spectroscopy for chemical analysis see X-ray photoelectron spectroscopy

electron tomography

Tomography using electrons as the source of

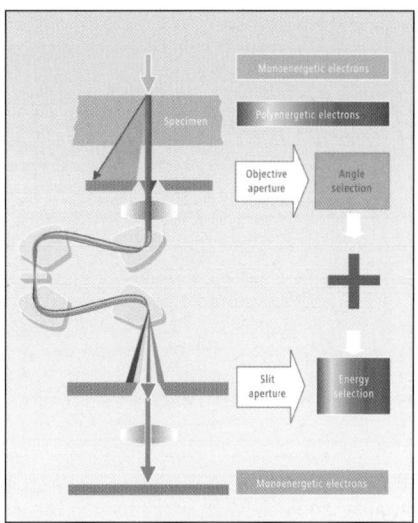

Principle of electron spectroscopic imaging with omega filter. Courtesy of Carl Zeiss SMT.

illumination. In electron tomography, a set of projection images obtained either from a tilt series of a single specimen, or from differently orientated multiple specimens, is obtained in bright- or darkfield modes, sometimes with the assistance of spectroscopic or holographic imaging, and the dataset is used to produce a computer-generated 3D reconstruction of the specimen.

electron trap
A magnetic trap that reduces entry of back-scattered electrons into an X-ray spectrometer. The trap lies between the collimator and the detector and comprises a pair of permanent magnets that deflect electrons.

electronvolt
A non-SI unit of electric potential difference or energy. Symbol eV. An electronvolt is the work done in moving an electron through a potential difference of one volt. 1 eV = 1.602×10^{-19} joules. Electronvolts are used to describe energy; the units of accelerating voltage are volts (V, kV).

electropolisher see **jet electropolishing device**
electropolishing see **electrolytic thinning**
electrostatic biprism see **electron biprism**

electrostatic energy analyzer
An electron spectrometer that uses electrostatic fields to disperse electrons.

electrostatic force microscopy
The use of a scanning probe microscope to map and measure charged domains in a specimen. In EFM a charged probe is scanned in non-contact mode across the specimen surface, detecting surface charges by changes in cantilever deflection or resonant frequency. A typical application is the mapping of electrostatic fields in electronic circuits.

electrostatic lens
An electron lens that uses electrically charged plates to deflect electrons. Electrostatic lenses respond more quickly to changes in lens excitation

than magnetic lenses (μs versus ms). A round electrostatic consists of several electrodes with a circular opening held at different voltages. The electrodes are often three in number though more electrodes provide greater flexibility. If the lens has no overall accelerating effect (the first and last electrodes are then held at the same potential), it is described as an einzel lens.

elemental compositional analysis
The investigation of the presence, concentration and spatial distribution of elements in a specimen.

elemental mapping see X-ray mapping

ellipsometry
The measurement of the change in the state of polarization of light by reflection from a surface.

elliptically polarized light
Light waves in which the electric field vector rotates 360° and changes its magnitude in each wavelength, describing an ellipse about the axis of propagation. Elliptically polarized light is produced when two linearly polarized waves with a relative phase shift of $n\lambda/8$ combine where n is an integer.

embedding
The surrounding and/or infiltration of a specimen with a matrix to facilitate sectioning; for example, paraffin wax embedding for histology, or resin embedding for transmission electron microscopy.

embedding medium
A matrix used to support specimens for subsequent processing or examination.

embedding mold
A cylindrical or flat mold for embedding resins, typically shaped to form a block ready for trimming or sectioning.

emission
The release of material or energy, such as photons, electrons or ions.

emission current see beam current

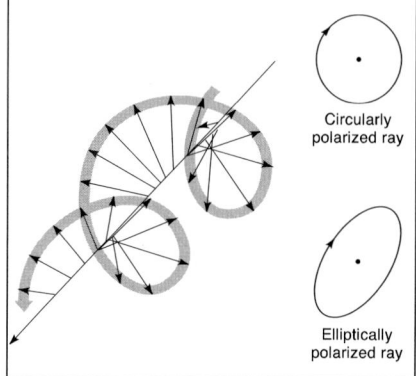

Circularly and elliptically polarized light. From D. Murphy © John Wiley and Sons, Inc.

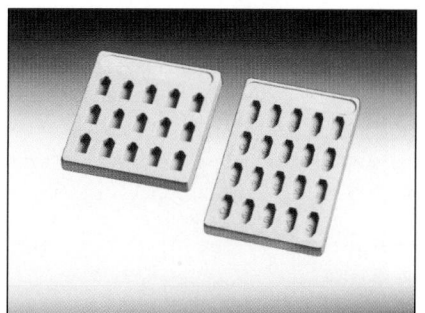

Molds for resin embedding. Courtesy of Agar Scientific.

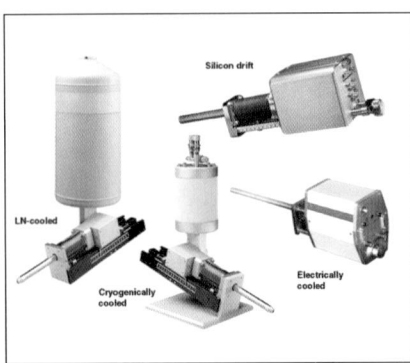

EDX spectrometers. Courtesy of Thermo Electron Corporation.

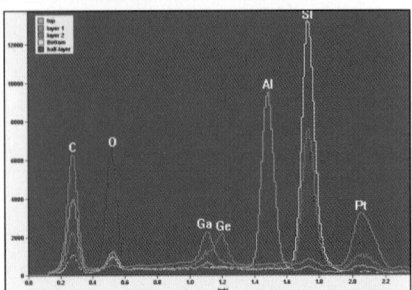

Energy-dispersive X-ray map (above) and X-ray spectrum (below) of cross-sectioned semiconductor. Courtesy of Thermo Electron Corporation.

emission maximum

The peak energy or wavelength in the emission spectrum of a fluorophore or excited species.

emission spectrum

A spectrum of the radiation emitted by a specimen or fluorophore.

emitter

A general term for a source of radiation or electrons, such as a filament.

empty magnification

Any magnification of a specimen or recorded image that does not reveal additional detail. In a light microscope, empty magnification occurs when the total magnification exceeds 500-1000 times the numerical aperture of the objective lens. Empty magnification of a digital image leads to pixelation.

emulsion see film emulsion

energy-dispersive spectrometer

A spectrometer that analyzes radiation by energy, typically by using a semiconductor detector.

energy-dispersive spectrometer

A spectrometer that analyzes radiation by its energy.

energy-dispersive spectroscopy

The use of an energy-dispersive spectrometer.

energy-dispersive X-ray spectrometer

A spectrometer that characterizes X-rays by energy. The key components of an energy-dispersive X-ray spectrometer are a semiconductor detector that generates a charge pulse proportional to the energy of an incident X-ray photon, processing electronics, a multichannel analyzer and a computer to control data acquisition, processing and display.

energy-dispersive X-ray spectroscopy

The identification of elements by analysis of the energies of the X-rays emitted by excitation of specimen atoms with a focused electron beam. Energy-dispersive X-ray spectroscopy of bulk specimens is performed in a scanning electron microscope or in a microprobe; thin sections are analyzed in an analytical electron microscope. Emitted X-rays have an energy that is characteristic of excited elements. The concentrations of most, except the lightest, elements can be detected in the 100 parts per million range. EDX has a lower resolution (peak FWHM ~140 eV) than WDX, but data acquisition is faster. The spatial resolution is a function of the electron probe size at the specimen; for thin sections this can be in the nanometer range with a field-emission gun.

energy-dispersive X-ray spectrum

A plot of intensity versus energy of emitted X-rays detected by energy-dispersive X-ray microanalysis.

energy filter

A spectrometer that disperses electrons according to energy, allowing selection of electrons of a specific energy range for imaging and analysis.

energy-filtered electron tomography

Electron tomography using an energy-filtering transmission electron microscope.

energy-filtered imaging see electron spectroscopic imaging

energy filtering

The use of an electron spectrometer to select a specific range of electron energies for image formation.

energy-filter(ing) transmission electron microscope

A transmission electron microscope equipped with an energy filter for electron energy-loss spectroscopy and electron spectroscopic imaging.

energy-filtering transmission electron microscopy

The use of an energy filtering transmission electron microscope.

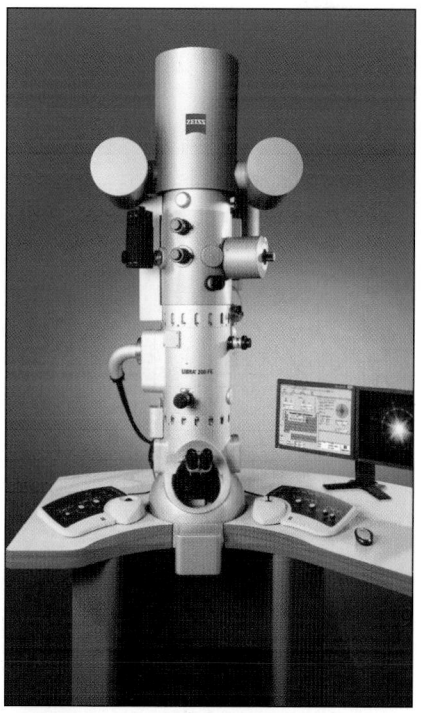

Energy-filtering transmission electron microscope. Courtesy of Carl Zeiss SMT.

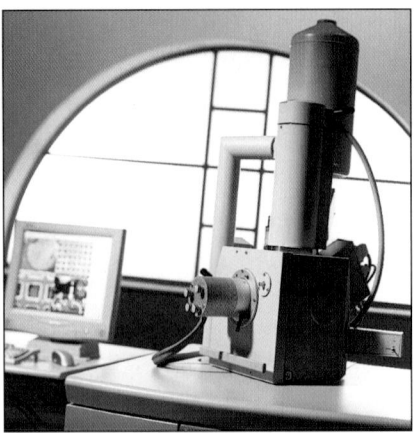

Environmental scanning electron micro-scope. Courtesy of FEI Company.

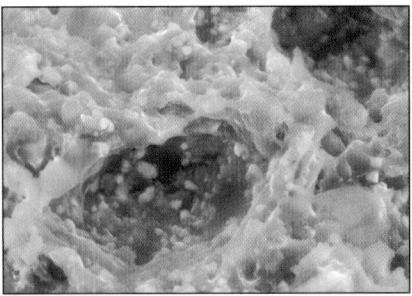

Wet bubble gum imaged in an environmental scanning electron microscope. Courtesy of FEI Company.

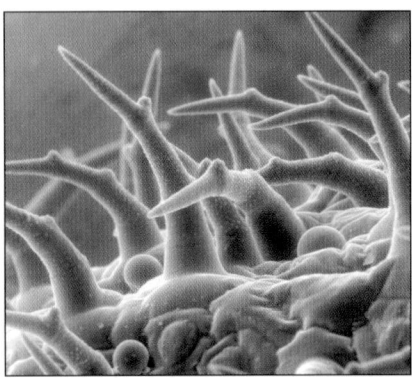

ESEM of unprocessed plant leaf. Courtesy of FEI Applications Laboratory.

energy-loss fine structure

The fluctuations in the intensity of an electron energy loss spectrum beyond an ionization edge. Energy-loss fine structure gives information on the bonding characteristics of the ionized atom.

energy-loss near-edge structure

The fluctuations in the intensity of an electron energy loss spectrum in a region up to 50 eV beyond an ionization edge. Energy loss near-edge structure reflects the unfilled density of states of the ionized atom, and gives information on atomic bonding characteristics.

energy-loss spectrum

A spectrum that shows the energy lost by radiation and charged particles transmitted or reflected by a specimen.

energy-selecting slit

An aperture (diaphragm) in the dispersion plane of an electron spectrometer, used to select electrons of a specific energy range.

enlarger see photographic enlarger

entrance pupil

The image of the aperture stop of an optical system viewed from an axial point in object space through any intervening lenses.

environmental scanning electron microscope

A brand-specific variable-pressure scanning electron microscope with a gaseous secondary-electron detector. An ESEM has a specimen chamber and column divided into pressure regions, separated by pressure-limiting apertures, allowing the examination of unprocessed and hydrated specimens in their native state at pressures up to 6,000 Pa, the minimum for liquid water.

environmental scanning electron microscopy

The use of an environmental scanning electron microscope.

environmental transmission electron microscopy

The use of a transmission electron microscope to investigate specimens enclosed in an environmental cell or chamber at pressures higher than that in the column.

enzyme cytochemistry

1. The localization of enzymes in cells by incubation in a specific substrate that produce a colored or dense reaction product in the presence of the enzyme.
2. The use of enzymes as probes in cytochemistry.

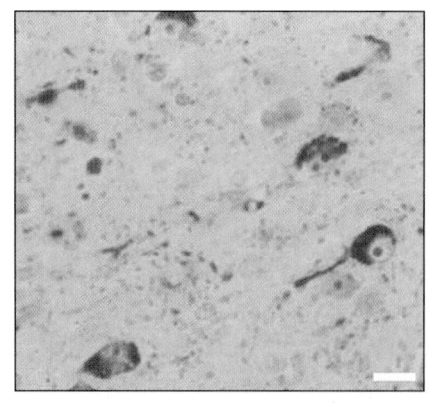

Enzyme cytochemistry of viral antigens in brain. *Microscopy and Analysis* 2004.

enzyme histochemistry see enzyme cytochemistry

epifluorescence microscope

A fluorescence microscope that uses epi-illumination. The key feature of an epifluorescence microscope is a filter cube, placed in the body tube, that selects excitation wavelengths, directs excitation light onto the specimen and transmits emission wavelengths to the eyepieces and imaging detectors.

epi-illumination

Illumination which is incident on an object on the side from which it is observed. Epi-illumination is the standard mode in confocal microscopy, fluorescence microscopy and reflected light microscopy.

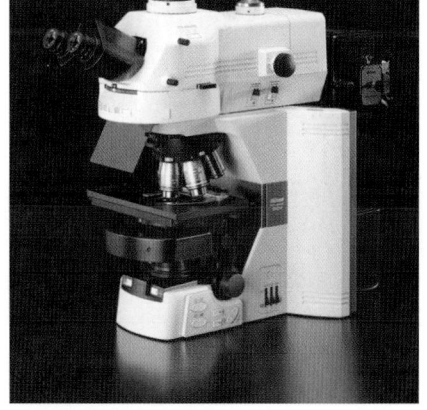

Epifluorescence microscope. Courtesy of Nikon.

epi-illuminator

A module of a light microscope that provides epi-illumination of the specimen through the objective. An epi-illuminator is inserted into the microscope tube and contains a light source, various optical elements and a beamsplitter to direct light down to the objective.

epi-objective

A light microscope objective designed for use with epi-illumination.

episcopic mode

The use of epi-illumination.

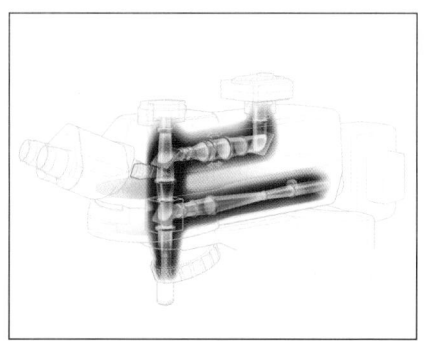

Optical pathways in epifluorescence microscope. Courtesy of Nikon.

epitaxy

The growth of a thin crystalline film of a material on a crystalline substrate by chemical vapor deposition or molecular beam deposition.

epitope

The region of an antigen to which an antibody binds. Epitopes are characterized by a specific conformation of atoms.

Epon

Proprietary name for a commonly used epoxy resin embedding medium

epoxy resin

A synthetic resin produced by polymerization of epoxides with phenols. Epoxy resins are cured in the presence of a hardener at temperatures between 5 and 150°C. Epoxy resins are widely used as embedding media for ultramicrotomy as they have low shrinkage and high mechanical strength.

equivalent focal length

The focal length of a multi-element optical system, having a value equivalent to that of a single lens.

E ray see extraordinary ray

erect image

An image having the same orientation to the observer as the specimen. Many optical systems produce an inverted intermediate or primary image; additional lenses or prisms in the optical train then rotate this image through 180°.

error signal mode see constant height mode

escape peak

A small peak in an X-ray spectrum often found below the normal peak for a particular element. Escape peaks are caused by loss of energy of primary X-rays due to secondary X-ray generation.

etalon

A device consisting of two closely spaced reflecting plates, used for the study of interference of waves.

etching

The removal of surface material by sublimation or by the action of chemicals, ionized particles or radiation.

eucentric

Correctly centered.

eucentric goniometer see goniometer

eucentric height

The position of the eucentric plane in the objective lens of a transmission electron microscope.

eucentric plane

The reference object plane in a transmission electron microscope. The eucentric plane is that plane in a goniometer at which tilting produces no lateral movement. The plane is adjusted with the Z control on the goniometer.

evanescent electromagnetic field see evanescent wave

evanescent wave

An electromagnetic wave whose amplitude decays exponentially. An evanescent wave is formed when light is totally internally reflected at an interface; some energy passes through the interface and propagates as an evanescent wave parallel to the interface. Evanescent waves form the basis of total internal-reflection fluorescence and attenuated infrared microscopies.

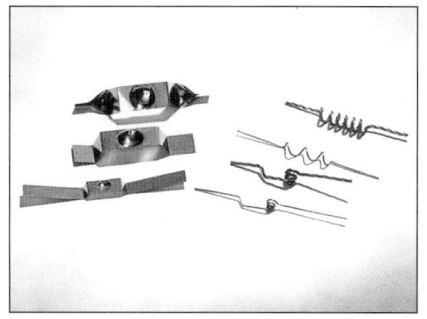

Evaporation baskets and boats. Courtesy of Agar Scientific.

evaporation

The change in state of a substance from liquid to gas at a temperature below its boiling point.

evaporation basket

A basket used to hold metal chips for evaporation in a coating unit, typically made of a high-melting point material such as tungsten wire (mp 3422°C).

evaporation boat

A container for powdered metals evaporated in a coating unit, often made of a high melting point material, e.g. molybdenum, platinum or tungsten.

evaporation source

The object from which material is evaporated in a coating unit, e.g. Pt filament or carbon rod.

evaporation unit

An apparatus for the coating of specimens and substrates using material evaporated from a target under vacuum.

Everhart-Thornley detector

A secondary electron detector widely used in electron microscopes. The Everhart-Thornley detector comprises a Faraday cage, a scintillator and a photomultiplier tube. Secondary electrons are attracted to the cage, which is biased with a positive potential of a few hundreds of volts, and then accelerated to the scintillator which is positively biased by an additional several thousand volts. The scintillator is connected by a light pipe to a photomultiplier tube.

Ewald sphere

A geometrical construction used in the interpretation of electron and X-ray diffraction patterns of crystalline materials. The diffraction patterns can be predicted by identifying the points of intersection of the Ewald sphere with the nodes of the reciprocal lattice, which is determined by the crystal lattice. The radius of an Ewald sphere is $1/\lambda$, so for short wavelengths its curvature can often be neglected. One diameter of the sphere represents the direct beam, and its surface passes through the origin of the reciprocal lattice. For any point in the reciprocal lattice that intersects the surface of the sphere, the set of planes that correspond to that point satisfy Bragg's law and will diffract strongly.

excitation

1. The process by which a molecule, atom or electron acquires energy that raises it from its ground state to a higher energy level. Upon relaxation to the ground state, the excited species releases energy, usually as a photon.
2. The application of electric current to a magnetic lens.

excitation balancer

A set of one or more interference filters placed in front of the light source in a fluorescence microscope that filter the emission from the source, removing wavelengths not required for fluorescence excitation and hence improving the efficiency of the filter cube.

excitation error

A measure of the deviation of a diffracted beam from the Bragg condition. The excitation error **s** is the distance from the Ewald sphere to a reciprocal lattice point measured along the normal to a thin crystalline slab.

exhaust pump

A type of vacuum pump that expels gases, such as rotary, diffusion or turbomolecular pumps.

exit pupil

The image of the aperture stop of an optical system viewed from image space through the intervening lenses.

exit wavefunction

The electron wavefunction evaluated on a two-dimensional plane immediately downstream of a thin sample in a transmission electron microscope. The exit wavefunction is generally a complex function with no simple relationship to sample potential apart from symmetry.

expanded-pupil eyepiece

A proprietary design of high-eyepoint eyepiece with a large exit pupil allowing greater freedom of eye and head movement by the observer. Expanded-pupil eyepieces incorporate multifaceted spinning transfer disks containing many minilenses.

exposure

The quantity of radiation that reaches a film or detector.

extended energy-loss fine structure

The ripples in the intensity of an electron energy loss spectrum stretching for several hundred eV after an ionization edge. Extended energy-loss fine structure arises because the electron ejected from the core shell during ionization is diffracted by surrounding atoms. Therefore the EXELFS intensity can be related to the radial distribution of atoms around the specific ionized atom.

extended-pressure scanning electron microscope

A variable-pressure SEM that operates at pressures in the range 1-750 Pa.

external diaphragm eyepiece

An eyepiece with a diaphragm on the object side of the lenses.

external reflection

The reflection of incident light when passing from a medium of low optical density (refractive index) to one of higher density.

extinction

The darkening of a monochromatically illuminated optically anisotropic material placed between crossed polars when a crystal vibration direction is parallel to the transmission axis of the polarizer.

extinction distance

The characteristic length of the diffraction vector **g**. Symbol: ξ_g; units: nm. The extinction distance is the periodicity in thickness of a Bragg beam under conditions where only two beams are excited within a crystal. It is observed as the period of the thickness fringes at the edge of a wedge in a dark- or bright-field two-beam image.

extinction factor

The ratio of the intensity of light transmitted by a polarizer and analyzer when their transmission axes are parallel, to the intensity when their axes are crossed.

extinction ratio see extinction factor

extraction anode

The first of the two anode plates in a field-emission gun whose role is to extract electrons from the filament. The extraction anode is

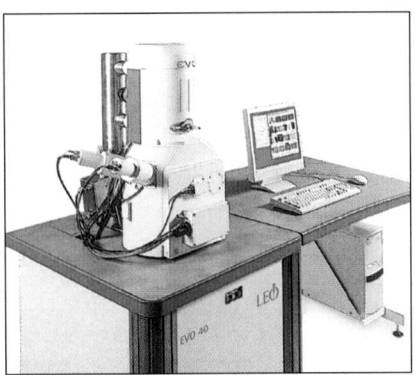

Extended variable pressure scanning electron microscope. Courtesy of Carl Zeiss SMT.

positively charged by several thousand volts relative to the filament.

extraction voltage

The voltage applied to a field-emission filament.

extractor plate see extraction anode

extraordinary ray

One of the two linearly polarized rays (the E and O rays), with mutually perpendicular vibrational planes, transmitted by a birefringent material. The E ray does not obey the normal law of refraction.

extreme infrared

Electromagnetic radiation with wavelengths in the range 15 μm to 1 mm.

extreme ultraviolet

Electromagnetic radiation with wavelengths in the range 4-200 nm.

extrinsic region see semiconductor detector

eye

The eye is a simple optical system that forms an inverted real image of an object on the retina. The information and parallax differences of the images received by each eye are interpreted by the brain as an erect three-dimensional image, i.e. stereoscopic vision. The optical power of the unaccommodated eye is around 60 diopters. The resolving power of the eye is ~1 arc minute, or about 1 mm at 3 m.

eyelash probe

A tool made by mounting an eyelash or fine bristle on a rod, typically used in microtomy for the manipulation of thin sections in a water trough or cryosections on a knife.

eyelens

The upper lens or lens system of a compound eyepiece that is closer to the eye.

eye lens

The flexible biconvex structure that forms the lens of the eye. The shape and hence the refractive

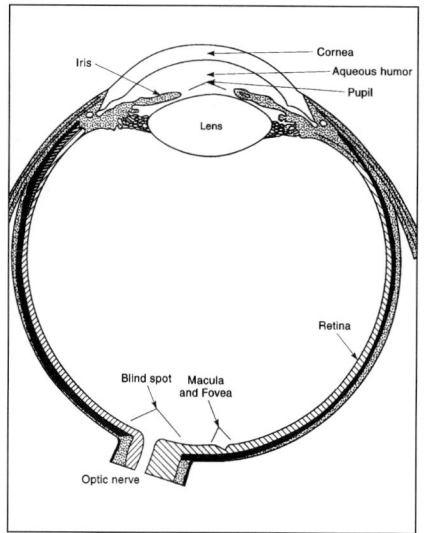

Diagram of human eye. From D. Murphy © John Wiley and Sons, Inc.

power of the eye lens is regulated by the ciliary muscles to allow accommodation.

eye-level riser

An extension to the tube of a light microscope used to raise the viewing head and eyepiece tubes to an ergonomic position.

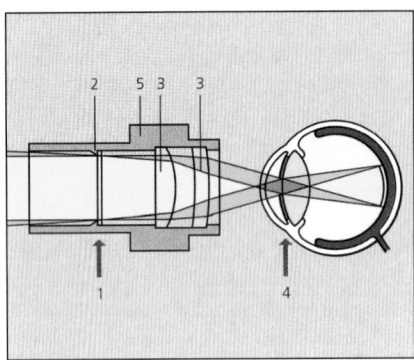

Eyepiece design and ray paths to eye. 1. Position of intermediate image and reticle. 2. Limit of field of view. 3. Lenses. 4. Position of Ramsden disk and observer's eye pupil. 5. Diopter adjuster. Courtesy of Carl Zeiss Ltd.

eyepiece

The uppermost lens or lens system of a compound microscope whose object is the intermediate image and which forms a magnified real or virtual image. The key features of an eyepiece are a front eyelens, additional multi-element lenses, a diaphragm that also supports a reticle (graticule), and a field lens. These components are housed in a barrel that may have a focusing or diopter adjustment mechanism and an eyetube fastening mechanism.

eyepiece, for glasses see **high-eyepoint eyepiece**
eyepiece graticule see **reticle**

eyepiece markings

The specifications engraved on the barrel of a light microscope eyepiece. Typically these markings provide information about: manufacturer; eyepiece name; magnification; eyepoint; field width; field of view number; aberration corrections; and diopter adjuster settings.

eyepiece micrometer see **reticle, micrometer screw eyepiece**

eyepiece tube

That part of the tube of a light microscope that carries the eyepiece. The eyepiece tubes can be moved together to adjust the interpupillary distance and tilted for ergonomic viewing.

Eyepieces. Courtesy of Carl Zeiss Ltd.

eyepoint see **exit pupil**

eye relief

The distance from the eyelens to the eyepoint of an eyepiece.

F

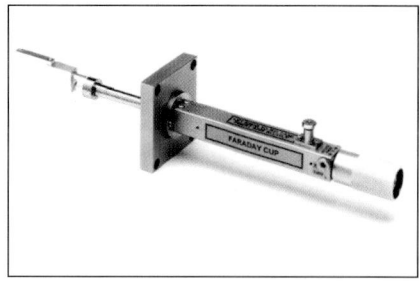

Faraday cup for probe current measurement in an SEM. Courtesy of Agar Scientific.

Fabry-Perot interferometer

A multiple-beam interferometer containing two closely spaced transparent plates with semi-transparent apposed surfaces. Light passing through the plates is multiply reflected from the apposed surfaces before emerging as parallel rays with differing optical pathlengths that are then focused to form an interference pattern.

Faraday cage

A cage of a conducting material such as copper wire mesh that excludes electric and magnetic fields. A Faraday cage may be installed in the walls of a room to protect electron microscopes and spectrometers from external magnetic fields.

Faraday cup

A device used to measure the beam current at the specimen in an electron microscope. A Faraday cup comprises an insulated metal box with an aperture to allow the electrons to enter but not escape; the current generated by the beam is read by a picoammeter. The cup may be built into the tip of a specimen holder.

Faraday cage. Courtesy of Kinetic Systems, Inc.

Faraday effect

The rotation of the plane of polarization of light passing through an isotropic material exposed to a magnetic field.

Faraday shield see **Faraday cage**

far-field diffraction see **Fraunhofer diffraction**

far-field optics

The use of optical systems where the object to lens distance is typically greater than one wavelength of light and so images are subject to the effects of diffraction.

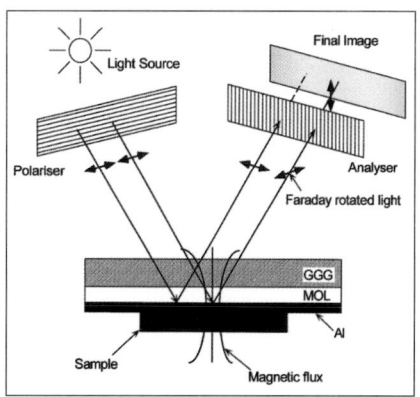

Faraday effect demonstrated by magneto-optical layer. *Microscopy and Analysis* 2001.

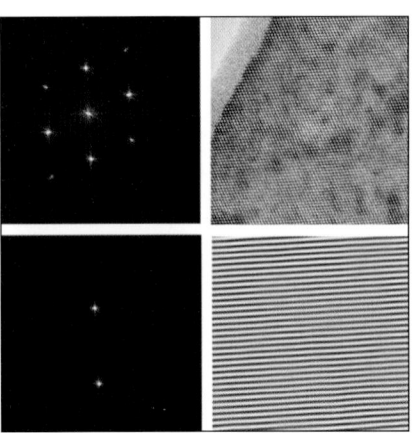

TEM image of monocrystalline GaAs (top left) and its fast Fourier transform (top right). By filtering the transform (bottom right) and applying an inverse FFT a single set of planes can be identified (bottom left). Courtesy of Soft Imaging System.

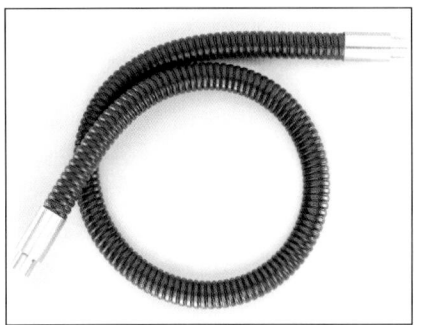

Fiber optic light guide. Courtesy of Stocker Yale Inc.

far infrared

Electromagnetic radiation with wavelengths in the range 6-15 μm.

far point

The object distance of the unaccommodated normal eye, usually located at least five meters from the eye.

far-sightedness see hypermetropia

far ultraviolet

Electromagnetic radiation with wavelengths in the range 200-300 nm.

fast axis

The axis of an optically anisotropic material along which light travels at highest velocity.

fast Fourier transform

An efficient method for applying a Fourier transform using a computer.

fast secondary electron see secondary electron

Fermat's principle

Fermat's principle states that light takes the least time in traveling between two points. In its modern form the principle may be restated as: a light ray traveling between two points must have an optical path length that is stationary with respect to variations of the path.

fiber-optic cable

A light conduit formed by a set of glass fibers. A fiber-optic cable contains many thousands of fibers, called cores, each 5-10 μm thick and surrounded with a reflective layer of cladding. Light passes through the cores by total internal reflection and emerges from the cable with uniform spatial intensity.

field cancelation see magnetic field cancelation

field curvature

An aberration of lenses that causes a plane object
to form a curved image plane so that only one part
of the image is in focus at one time. Field
curvature depends linearly on angle and quadratically on off-axis distance at the object plane.

field diameter

The maximum diameter of the area of a specimen
that can be viewed in a light microscope. It is
equal to the field number divided by the
magnification of the objective.

field diaphragm

The diaphragm that limits the area of the object
that can be imaged by an optical system. In a light
microscope, field diaphragms are placed in front
of the light source and in the eyepiece.

field electron microscope see field-emission microscope

field emission

The emission of electrons from a sharply tipped
metal source in the presence of an electric field.
The field must exceed the work function of the
source to allow electrons to tunnel through a
surface potential barrier and escape.

field emitter

A filament that releases electrons by the cold or
Schottky field-emission mechanism.

field-effect transistor

A pre-amplifier used in an X-ray spectrometer that
converts charge pulses into voltages, producing a
staircase voltage waveform where the height of
each step is proportional to photon energy. The
FET is reset before saturation by applying an
electrical or optical signal.

field-emission gun

An electron gun that uses field-emitters to form an
electron beam. The major components of a field-
emission gun are: a field emitter; a supressor cap
(Wehnelt); two anode plates, the first providing a
positive extraction voltage of several thousand
volts, and the second the accelerating voltage.

Field-emission guns require an ultrahigh vacuum and may have an electrostatic or magnetic gun lens and a monochromator.

field-emission gun scanning electron microscope

A scanning electron microscope equipped with a field-emission gun.

field-emission gun transmission electron microscope

A transmission electron microscope equipped with a field-emission gun.

field-emission microscope

A microscope that forms images of specimens that emit electrons by cold or thermal field emission. A needle-shaped specimen is placed opposite a fluorescent screen or detector in a high vacuum chamber. A high negative voltage applied to the specimen extracts electrons forming an image on the screen that identifies planes of atoms at the specimen surface.

field-emission source see field emitter

field-ion microscope

A microscope that forms images of metal and alloy specimens using field ionization. A cooled (10-150 K) needle-shaped specimen with a high (~30 kV) positive potential is placed in a vacuum chamber containing a low pressure inert imaging gas (e.g. He or Ar); as gas molecules approach the tip they are ionized and then accelerated towards a detector forming an image representative of the atomic structure of the tip surface. Atoms in the specimen may also be desorbed (evaporated) by ionization and contribute to image formation.

field-ionization microscope see field-ion microscope

field iris see illuminated field diaphragm

field lens

The lens or lens system of an eyepiece closest to the object.

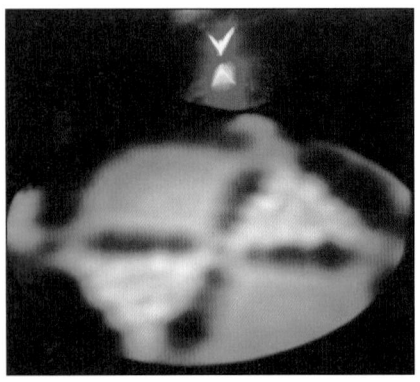

Field emission microscope images of tungsten crystal. *Microscopy and Analysis* 2003.

field microscope

A microscope that can be used outdoors. Field microscopes are of compact design with few accessories, are easily transportable, and have a battery operated power supply or use the sun as a light source.

field number

The diameter of the field of view at the intermediate image plane of a light microscope. The field number is the diameter in millimeters of the diaphragm (or its image) in the eyepiece.

field of video

One vertical sweep of a raster scan. In a 2:1 interlaced video signal, two fields sweeping alternate lines produce one full video frame.

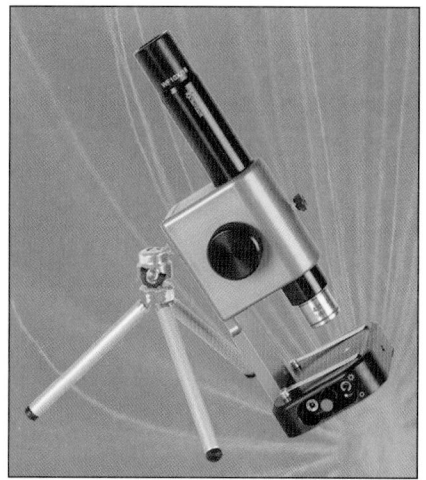

Field microscope. Courtesy of Richardson Technologies.

field of view

The area of an image that falls upon the retina of the eye or a detector.

field of view number see field number

field plane

The object plane and all planes conjugate with the object plane.

field stop see field diaphragm

filament

A collective term for thin metal elements that emit light or electrons.

filament heating current

The current flowing through and heating the filament of an electron gun. In a self-biasing gun, the filament current is normally set to a value to achieve filament saturation.

filaments, for electron guns see tungsten filament, lanthanum hexaboride filament

filament saturation

The optimization of the emission of the filament in a self-biasing electron gun. For any setting of the gun bias resistor, there is a point on the sigmoidal emission current versus filament temperature plot where emission stabilizes or reaches saturation.

filar eyepiece see micrometer screw eyepiece

filar micrometer see micrometer screw eyepiece

film
1. A thin sheet of plastic covered with photographic emulsion. 2. A thin layer of material on a substrate.

film cassette
A light-proof container for film. Roll film is rewound into the cassette after exposure. In a transmission electron microscope, the film plate cassette is usually in two parts: one for unexposed film and one for exposed film.

film desiccant
An anhydrous material placed in a film storage container to desiccate film; e.g. silica gel, molecular sieve.

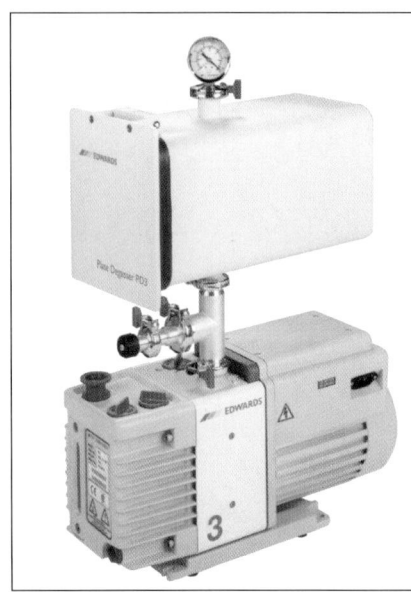

Film vacuum desiccator for EM roll and sheet film. Courtesy of Agar Scientific.

film desiccator
A vacuum chamber for the degassing and drying of photographic film.

film dryer
A warm air cabinet for drying photographic film and paper after chemical processing and washing.

film emulsion
A thin film of gelatin containing silver halide grains which are turned to silver metal after exposure to radiation and chemical development.

film exposure see exposure

film, for TEM
Film for transmission electron microscope cameras is available in a range of sizes and speeds. TEM film typically has a very fine grain emulsion on a 0.17-mm thick polyester base. Nominal film speed of standard film is ISO 25; faster film (>ISO 50) is often used for low image intensity

applications such as HREM and cryo-EM. Standard film sizes are 8.25 x 10.16 cm and 6.5 x 9.0 cm. A notch in the upper right corner identifies that the emulsion side is uppermost.

film plate holder

A metal tray that carries a sheet of film in the film cassette of a transmission electron microscope.

film speed see ISO film rating

film-thickness meter

A device that measures the amount of material deposited on a substrate or specimen during a coating procedure. The meter can be qualitative, e.g. an oil droplet on porcelain or a piece of filter paper, or quantitative, e.g. a quartz crystal monitor.

film vacuum chamber see film desiccator

filter

A device or mechanism that selectively suppresses or separates components of a signal, specimen, image, or radiation.

Filter cubes in the epi-illumination pathway
Courtesy of Nikon.

filter block see filter cube

filter cube

A cubic housing for filters and mirrors used in epi-illumination modes in a light microscope, e.g. reflected light microscopy and epifluorescence microscopy. A filter cube typically has three ports, facing the light source, specimen and eyepieces. The center of the block contains a 45° inclined semi-silvered or dichroic mirror that reflects light onto the specimen and transmits reflected or emitted light to the eyepieces and detectors.

filter cube holder

A device that holds several filter cubes and can be rotated or moved laterally to select a filter set.

Filter block slider. Courtesy of Carl Zeiss Ltd.

filter kernel

An image filter that changes the value of each pixel in a digital image using mathematical operations specified by the kernel matrix. For example, a 3 x 3 median filter replaces the value of a pixel with the median value of a 3 x 3 matrix of pixels centered on that pixel.

filter slider see filter cube holder

filter tray

A swing-out receptacle for optical devices such as filters, relay lenses and polarizers, located between the source and condenser of a light microscope.

filter wheel

A wheel containing several ports for optical filters. A filter wheel is placed in front of the light source of a light microscope and is motor driven for rapid switching of filters.

finder grid

An electron microscope support grid with fiducial markers, such as numbers or letters, enabling (re)location of a specimen or region of interest.

fine structure see ultrastructure

FireWire

A proprietary name for a IEEE standard 1394 high-speed serial databus connection enabling digital data from a camera and other imaging devices to be transferred into computer memory. Other manufacturers use the names i.Link or Lynx to describe similar 1394 products.

first-order Laue zone see Laue zones

first-order optics

The study of perfect optical systems, free of any aberrations.

first-order red plate see full-wave plate

fix, fixer

A solution of sodium thiosulfate that reacts with the unexposed silver bromide in a film emulsion, converting it to silver sodium thiosulfite, so making the film insensitive to light.

fixation

The stabilization of the molecular structure of specimen. Fixation protects or reduces damage to

Filter wheel. Courtesy of Prior Scientific Instruments Ltd.

the specimen during subsequent specimen processing or microscopical examination.

fixative
Any agent that stabilizes a material or specimen.

Fizeau fringes
The irregular interference fringes produced by a non-uniform film between two reflecting surfaces, such as are found in a soap film.

flashing
The heating of a cold field-emission filament to remove adsorbed contaminants such as residual gases that may impair its performance.

flashing stereo display
A stereo display that alternately displays each image of a stereo pair at high speed.

flat-field objective
A light microscope objective corrected to produce a flat field of view.

flexibilizer see plasticizer

flint glass
A family of optical glasses with high refraction and dispersion, with Abbe number lower than 55.

flow cytometer
An instrument for measuring the fluorescence and light scattering properties of populations of particles or cells in suspension. A flow cytometer passes a stream of fluid containing fluorescently labeled cells or particles through a narrow flow chamber which is illuminated by a focused beam of light. Fluorescence and/or light scattering is measured by photomultipliers. The particles can be sorted after passing through the chamber by breaking the stream into droplets which are charged and deflected by electrostatic plates into separate reservoirs.

flow cytometry
The use of a flow cytometer.

fluctuation electron microscopy

An application of darkfield transmission electron microscopy that reveals information about the medium-range order in an amorphous specimen by statistical analysis of scattering from small volumes.

fluor

An abbreviation for fluorochrome, fluorophore or fluorescent dye.

fluorescence

Luminescence that occurs on a timescale of picoseconds or several (<10) nanoseconds. The expression fluorescence is commonly used to describe the emission of light by fluorophores and X-ray emission under X-ray excitation. Electron microscopists do not normally use this term for electron-induced X-ray emission.

fluorescence anisotropy

The dependence of fluorophore excitation on the state of polarization of incident light. Anisotropic fluorophores emit polarized light.

fluorescence correction

A correction factor used in quantitative X-ray microanalysis that accounts for X-ray fluorescence within the specimen.

fluorescence correlation spectroscopy

A technique in fluorescence microscopy for the measurement of diffusion coefficients and the study of molecular interactions. Fluorescence correlation spectroscopy measures the temporal fluctuation of fluorescence from small numbers of fluorophores in very small volumes (~1 femtoliter) which can be caused by diffusion or modification of the fluors by their environment.

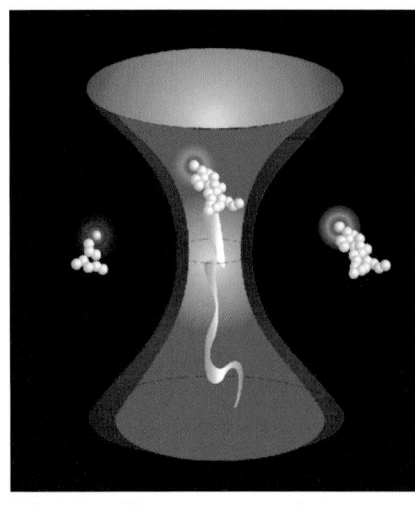

Principle of fluorescence correlation spectroscopy. Fluors are excited in a small focal volume. Courtesy of Leica Microsystems.

fluorescence filter

An optical filter that is used to select or block those wavelengths of light that excite or are emitted from a fluorophore.

fluorescence filter cube

A filter cube used in epifluorescence microscopy. A fluorescence filter cube typically contains filters to select excitation and emission wavelengths and a 45° dichroic mirror to reflect light onto the specimen and transmit light to the eyepieces and detectors. Several filter cubes with different specifications may be placed in a sliding or rotatable holder to allow selection of a cube appropriate for a specific fluorophore.

fluorescence lifetime

The average time that a population of fluorescent molecules simultaneously excited spends in the excited state before returning exponentially to ground state and emitting photons. Fluorescence lifetimes are typically on the order of 1-10 ns.

fluorescence-lifetime imaging microscopy

A form of fluorescence microscopy in which image contrast reveals the fluorescence lifetimes of fluorophores. FLIM is typically used to probe intracellular compartments of living cells as fluorescence lifetime is sensitive to ionic environment. Two modes are used: time-domain FLIM and frequency-domain FLIM.

fluorescence localization after photobleaching

A technique used to study the dynamics of molecules in living cells by selective photobleaching and fluorescence ratioing. In FLAP molecules are labeled with two differently colored fluorophores, one to be photobleached, the other to act as a reference. After photo-bleaching a selected area of a cell, difference images are produced by subtracting images of each fluorophore's distribution in the cell, allowing the relocation of molecules containing bleached fluorophore to be tracked.

fluorescence loss in photobleaching

A fluorescence technique to study the diffusion or trafficking of molecules between compartments in living cells. Molecules that reside in two compartments are labeled with fluorophores. One fluorescent compartment is then repeatedly photobleached; if fluorophores are moving between the two compartments, the second one will gradually lose fluorescence.

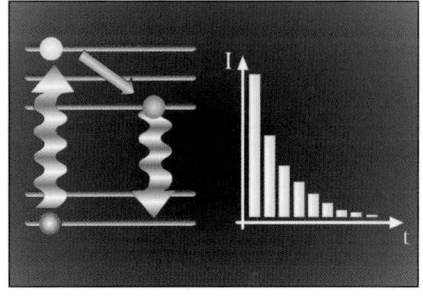

Principle of fluorescence lifetime.
Courtesy of Leica Microsystems.

fluorescence microscope

A light microscope with accessories for the examination of fluorescent specimens. The key accessories required for fluorescence microscopy are a source of excitation light, such as an arc lamp or a laser, filters to select excitation and emission wavelengths, and fluorescence objectives. The standard configuration is that of an epi-fluorescence microscope, but transmitted fluorescence microscopes are available.

fluorescence microscopy

The use of a fluorescence microscope to excite fluorescent molecules and use the emission wavelengths to form an image.

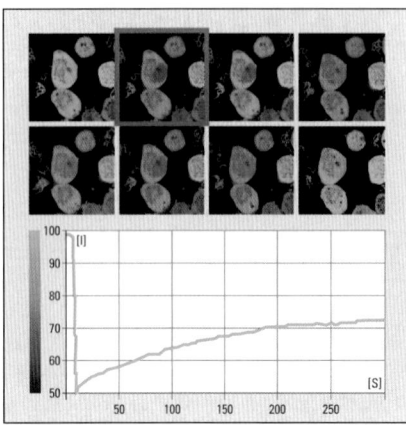

Fluorescence recovery after photobleaching. Time series of images showing the recovery (top) and graph of intensity vs time (bottom). Courtesy of Leica Microsystems.

fluorescence photobleaching recovery see fluorescence recovery after photobleaching

fluorescence quantum efficiency see quantum efficiency

fluorescence ratioing

A technique in fluorescence microscopy that compares the images of fluorophores at different excitation or emission wavelengths. Fluorescence ratioing can be performed with two or more fluorophores, or with a single ratiometric probe.

fluorescence recovery after photobleaching

A fluorescence technique to study the dynamics and measure diffusion coefficients of molecules in living cells by photobleaching. A defined region of a fluorescently labeled cell is exposed to an intense pulse of laser light to photobleach all fluorophores; the laser is then attenuated and used to monitor the recovery of fluorescence in the bleached region by diffusion of fluorophores from other regions of the cell.

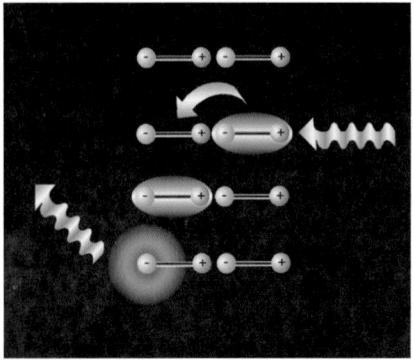

Principle of fluorescence resonance energy transfer. Courtesy of Leica Microsystems.

fluorescence resonance energy transfer

A technique to study the interaction of two fluorescently labeled molecules, e.g. receptor and ligand. Fluorescence resonance energy transfer occurs when the emission spectrum of one fluorophore (the donor) overlaps the excitation spectrum of a second fluorophore (the acceptor); when the donor is excited it transfers energy to the acceptor without emission of a photon (i.e. non-radiatively) through dipole-dipole interactions and the acceptor uses that energy for excitation and emission. The probability that energy transfer will occur depends on the sixth power of the distance between the fluorophores and is most efficient when the donor and acceptor lie within the Förster radius, typically 1-10 nm.

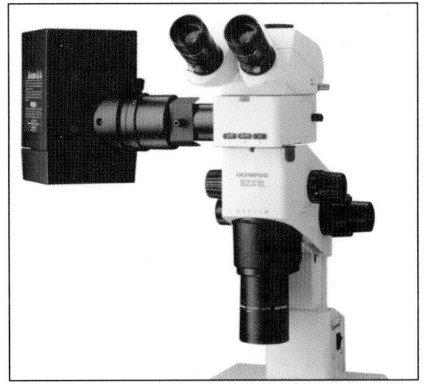

Fluorescence stereomicroscope. Courtesy of Olympus.

fluorescence stereomicroscope
A stereomicroscope equipped with an epi-illuminator and a filter cube for fluorescence microscopy, typically used for examining large fluorescent specimens.

fluorescence yield
1. The fraction of ionization events that result in the emission of a characteristic X-ray. The fluorescence yield decreases rapidly with atomic number, reaching $<10^{-3}$ for carbon.
2. Synonym for fluorescence quantum efficiency.

fluorescent cell sorting see flow cytometry

fluorescent dye see fluorophore

fluorescent in-situ hybridization
The use of fluorescent probes in in-situ hybridization.

fluorescent proteins
A family of fluorescent molecules naturally present in luminescent marine organisms, e.g. the jellyfish *Aequorea victoria* and the sea anenome *Discoma striata*. There are many types of fluorescent proteins, both natural and genetically engineered, with differing excitation and emission maxima covering the whole spectrum of visible light. Commonly used FP include: blue (e_{max} 440 nm), cyan (e_{max} 477 nm), green (e_{max} 508 nm), yellow (e_{max} 527 nm), and red (e_{max} 583 nm). In GFP the fluorophore is a tripeptide ring (serine-tyrosine-glycine) which is modified in variants to produce different emission maxima.

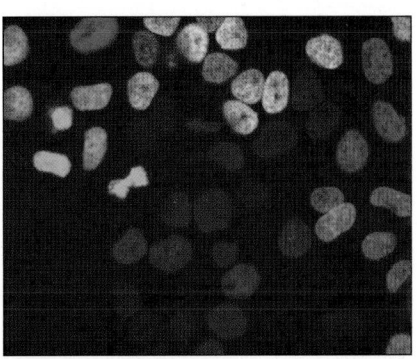

Fluorescent proteins: Cell nuclei labeled with GFP, YFP and RFP. Courtesy of Nikon.

fluorescent speckle microscopy

A fluorescence technique to study the dynamics of macromolecular assemblies using low levels of fluorophores. Polymeric structures such as the microtubules in living cells are labeled with below saturating levels of fluorescent macromolecules resulting in a speckled appearance in a fluorescence image. The speckles can be treated as spots which can then be tracked individually to follow the movement of the labeled structure and the dynamics of its components.

fluorimetry

The measurement of fluorescence.

fluorite

Crystalline calcium fluoride (CaF_2). Refractive index at 587 nm = 1.434.

fluorite objective see semiapochromat

fluorochrome

A stain that has the property of fluorescence.

fluorophore

The chemical domain of a stain that exhibits fluorescence. When a photon excites a fluorophore its electronic and vibrational energy levels are raised from ground state (S0) to higher states (S1, S2, etc). Upon relaxation to the ground state the fluorophore releases energy in the form of a photon. Energy is lost during this process so the emitted photon has a lower frequency than the exciting photon, a phenomenon called the Stokes shift.

fluorospar see fluorite

fly-eye lens

A lens formed from a 2D array of smaller lenses, like that of a fly-eye, that transmits uniform illumination.

F number

A measure of the light-gathering power of a lens. Symbol f/#, where # is a number. The F number is the ratio of focal length to effective aperture.

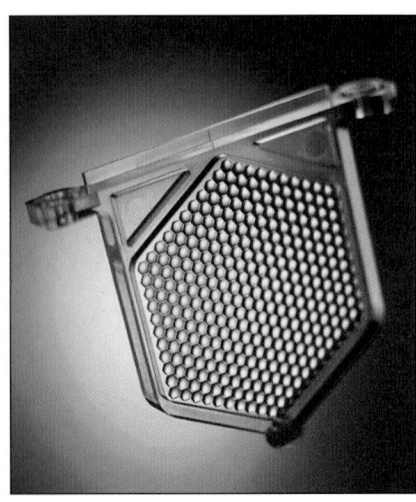

Fly-eye lens. Courtesy of Nikon.

focal length

The distance on the optical axis between a principal plane and its respective focal plane.

focal plane

1. The surface connecting all points at which beams of parallel rays passing through a converging lens are brought to a focus.
2. A common term for that plane in a thick specimen which is in focus in a microscope.

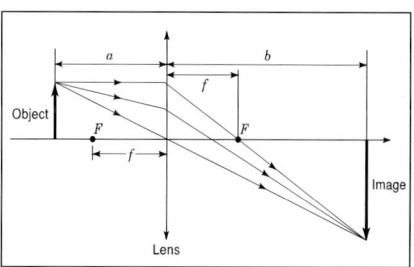

Geometrical optics of a thin lens showing the focal lengths f, focal points F, and object a and image b distances. From D. Murphy © John Wiley and Sons, Inc.

focal points

The points on the optical axis intersected by the focal planes.

focus

1. The focal point of an optical device.
2. The state of a sharp image.
3. The procedure used to obtain a sharp image.

focused ion beam

A focused beam of ions, typically provided by a liquid metal source, used to mill specimens to reveal underlying layers for microscopical analysis and for nanofabrication.

focused ion-beam microscope

A microscope that uses a focused ion beam to mill and sputter the specimen and forms images using secondary ions. A focused ion-beam microscope may also have secondary or backscattered electron imaging capability.

focused ion-beam milling

Milling with an ion beam focused to a diameter of >5 nm.

focused ion-beam scanning electron microscope

A scanning electron microscope containing a focused ion-beam system.

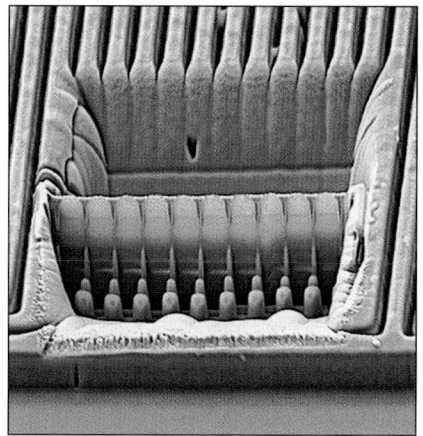

Focused ion beam milling of materials. Courtesy of Carl Zeiss SMT.

focused ion-beam transmission electron microscopy

The combined use of a focused-ion beam instrument to shape a specimen and a transmission electron microscope to image and analyze it.

focusing aid, for enlarger

A magnifying device used to focus negatives on an enlarger, comprising a mirror and a simple eyepiece.

focusing circle see **Rowland circle**

focusing screen

A small screen used as a focusing aid. In a transmission electron microscope the focusing screen can be swung into the optical axis manually or automatically and used in conjunction with the binoculars for fine focusing of the image.

focus wobbler

A wobbler that aids focusing of the image in scanning and transmission electron microscopes. The focus wobbler alternates the current in the beam tilt coils so translating an under- or over-focused image across the viewing screen, creating a 'wobbling' or vibrating image that is only stable when the point of focus is reached.

fogging

The unintentional exposure of photographic film by stray light, electrons or X-rays.

foil

A term commonly used by materials scientists to describe a thin section of a specimen.

folding grid

A composite electron microscope support grid that can be folded around a specimen and optionally latched.

force constant see **spring constant**

force modulation microscopy

The use of an atomic force microscope to measure and map the mechanical or viscoelastic properties of a specimen. In FFM the probe is placed in contact with the specimen surface with a constant cantilever deflection. A vertical oscillation is then applied to the tip (or specimen); the change in cantilever oscillation amplitude during scanning is a measure of the relative stiffness or elasticity of the specimen surface.

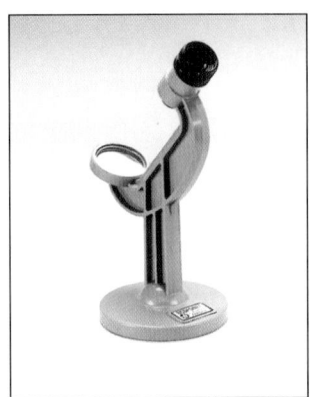

Focusing aid for photographic enlarger. Courtesy of Agar Scientific.

Folding grid. Courtesy of Agar Scientific.

foreline trap

A device placed between a rotary pump and an evacuated chamber to prevent backstreaming of oil.

formaldehyde

HCHO. A chemical fixative suitable for all biological specimens and widely used in histology. Formaldehyde (a gas) reacts with water to form methylene glycol (hydrate). The aldehyde group binds to amine groups in proteins forming a hydroxymethyl group ($-CH_2OH$) that then reacts with hydrogen forming a methylene bridge ($-CH_2-$). Formaldehyde penetrates tissues rapidly, but most crosslinks are reversible.

formalin

A solution of 40% formaldehyde widely used as a chemical fixative for histology.

Formvar

Proprietary name for polyvinyl formal/formate, a material used to make plastic support films.

Förster critical distance see Förster radius

Förster radius

The separation distance between acceptor and donor molecules that allows 50% efficiency of non-radiative transfer of energy in fluorescence resonance energy transfer.

Förster resonance energy transfer see fluorescence resonance energy transfer

forward scattering

Scattering at an angle less than 90° with respect to the incident beam.

Foucault imaging

In Lorentz electron microscopy, the generation of an image using electrons that have been deflected by magnetic domains in the specimen. Foucault imaging is analogous to darkfield imaging; the objective aperture is displaced laterally to exclude the direct beam allowing only deflected electrons to pass.

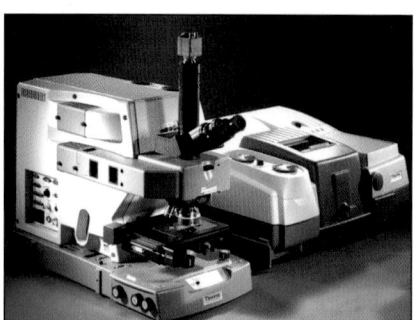

Fourier-transform infrared microscope.
Courtesy of Thermo Electron Corporation.

Fourier transform

A mathematical representation of an image or signal as the sum of a series of two-dimensional sine waves, each representing the spatial frequency of details in the image

Fourier-transform infrared microscope

A light microscope with accessories for Fourier-transform infrared spectroscopy. The key features of a FT-IR microscope are a source of infrared radiation (a laser or arc lamp), an interferometer, infrared objectives, and a CCD or point detector. The interferometer produces an interferogram, encoding the spectrum of the IR source, which is used to illuminate the specimen. Transmitted or reflected light is received by a detector that sends its output to a computer which decodes the interferogram using fast Fourier transform processing to produce an infrared spectrum from any point in the specimen.

Fourier-transform infrared microscopy

The use of a Fourier-transform infrared microscope.

Fourier-transform infrared Raman microscope

A microscope equipped with accessories for FT-infrared and Raman spectroscopy.

Fourier-transform infrared spectroscopy

A form of infrared spectroscopy in which there is simultaneous detection of all wavelengths of infrared absorption using Fourier analysis.

Fourier-transform infrared spectrum see infrared spectrum

Fourier-transform Raman spectroscopy

A form of Raman spectroscopy in which there is simultaneous detection of all wavelengths of Raman scattered light using Fourier analysis.

Fourier-transform Raman microscope

A Raman microscope in which light scattered by the specimen is passed through an interferometer before reaching the detector. The interferometer produces an interferogram that encodes all the frequencies of Raman scattered light from which the individual frequencies (wavelengths) can be rapidly decoded by Fourier transform processing. FT-Raman microscopy is suited to samples that fluoresce.

four-pi confocal microscope

A confocal microscope with coherent illumination provided by two opposed high-NA objectives. A 4pi microscope produces a spherical focal spot, doubling angular aperture to close to the 4π solid angle (hence the name), with a significant improvement in axial resolution.

fovea

A 0.5 mm-diameter pit at the center of the macula containing only cones. The fovea is used for acute vision.

frame

One complete image formed by a camera or monitor. A frame of video can be formed by a single field or by interlacing two or more separate fields. In a CCD a frame is formed by a full parallel shift.

frame averaging

The averaging or two or more consecutive frames. Frame averaging is used in imaging devices to reduce random noise.

frame grabber

An electronic device that digitizes a frame of analog video and stores it in a frame buffer. Frame grabbers usually have additional image processing capabilities such as frame integration, frame averaging and background subtraction.

frame integration

The addition of two or more consecutive frames or images. Frame integration is used in imaging devices to enhance contrast and reduce noise.

frame rate

The number of frames displayed per second. Standard broadcast video frame rates are 50 fps (PAL) and 60 fps (NTSC).

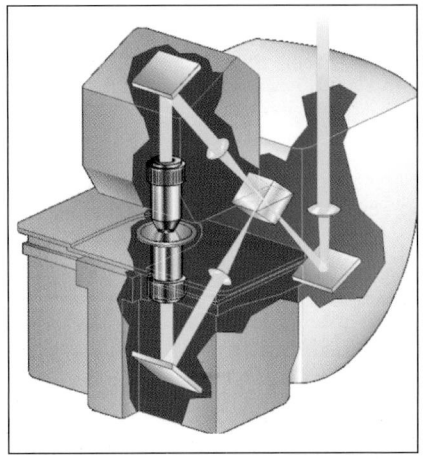

Optical pathways in a 4pi confocal microscope. Courtesy of Leica Microsystems.

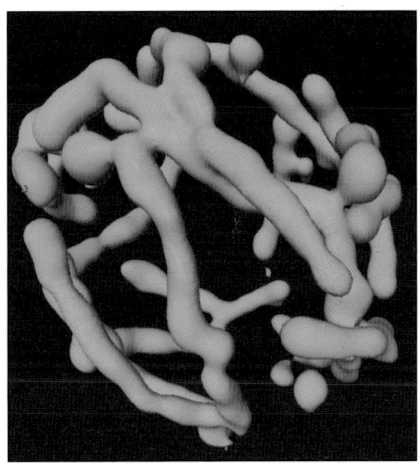

4pi confocal microscope image of mitochondrial matrix in living yeast (labeled with EGPF). Courtesy of Leica Microsystems.

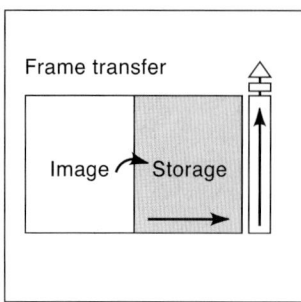

Imaging and storage areas of
frame-transfer CCD. From D.
Murphy © John Wiley and
Sons, Inc.

frame-transfer charged-coupled device

A type of charge coupled-device with a parallel
register divided into two halves or arrays, one for
imaging and a masked area for charge storage and
read out. In operation, the imaging array acquires
charges which are then transferred to the storage
array; charges are read out while the imaging array
is accumulating new charges. Frame-transfer
CCDs operate without a shutter and produce a
video-rate signal.

Fraunhofer diffraction

Diffraction in which the source and the image are
at infinite distances from the object, so the
wavefronts can be treated as planar.

Fraunhofer lines

Element-specific absorption lines in the spectrum
of solar radiation. Examples: the blue F line at
486.1 nm due to hydrogen, the yellow D_3 (d) line
at 587.5 nm due to sodium, and the red C line at
656.3 nm due to hydrogen. Fraunhofer lines are
used as reference wavelengths in the tracing of
paraxial rays and in the measurement of the
refractive index and dispersion of lens glasses.

free-break method

A standard method used to prepare glass knives
from glass squares. In the free-break method a
glass square is scored at a slightly oblique angle to
a diagonal allowing the fracture plane to run from
the ends of the score line to the closest edge,
producing an extremely sharp blade.

free viewing see infinity viewing

freeze drying

The dehydration of a frozen specimen by
sublimation, usually at low pressure ($\sim 10^{-3}$ Pa) and
low temperature ($<-60°C$). A technique used to
prepare dry specimens for scanning electron
microscopy and X-ray microanalysis.

CryoSEM image of freeze-fractured yeast
cells. Courtesy of FEI Applications
Laboratory.

freeze etching

The sublimation of ice at low temperature and low
pressure.

freeze fracture

The fracturing of a frozen specimen, usually at very low temperatures (~-196°C). The freeze-fracture technique is typically used to gain access to the interior of a specimen for structural or cytochemical studies. Fracture planes tend to follow paths of least resistance: in cells and organelles the planes often pass between bilayers, allowing the study of membrane lipids and proteins.

freeze-fracture device

An apparatus for specimen preparation by the freeze fracture technique. The key features of a freeze-fracture device are a high-vacuum chamber, a cold stage, a microtome knife for cutting and fracturing, and evaporators for specimen coating. A cryofixed specimen is placed on the cold stage and fractured with a knife blade at -180°C under high vacuum. The exposed surface can then be etched by raising the stage to ~100°C. A replica is then made of the exposed specimen usually using electron beam evaporation of Pt followed by a stabilizing coating of carbon. The specimen is then thawed and the replica released from the specimen for examination by transmission electron microscopy.

freeze substitution

The replacement of the ice in a cryofixed specimen with an organic solvent at low temperature.

freeze-substitution device

A device for the automated freeze substitution of specimens. A freeze substitution device has a cryochamber for the cryofixed specimen and the substitution solvent and post-fixative. After cryofixation the specimen is placed in substitution solvent, e.g. acetone, ethanol or methanol, which may contain a fixative, e.g. glutaraldehyde, osmium tetroxide or uranyl acetate, at very low temperature (<80°C) for several hours. Following substitution the now fixed specimen is warmed to room temperature and processed for microscopy.

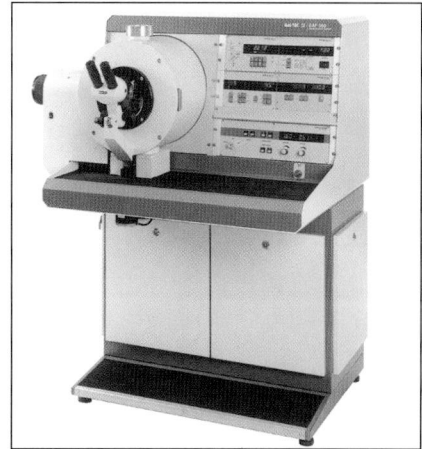

Freeze fracture device. Courtesy of BAL-TEC AG.

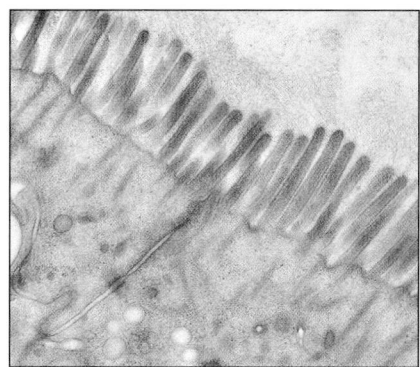

TEM image of cryofixed, freeze-substituted cells. Courtesy of BAL-TEC AG.

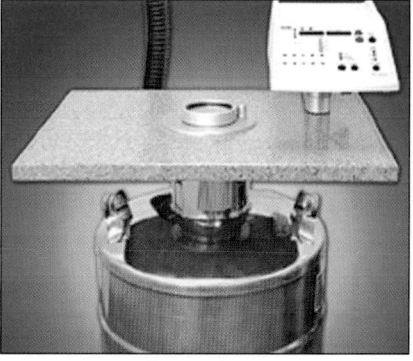

Freeze substitution device. Courtesy of Boeckeler Instruments, Inc.

freezing pliers

A pair of pliers equipped with polished copper jaws that can be cooled in liquid nitrogen and used for for cryofixation of small specimens and in-vivo tissues.

frequency

The number of cycles or objects per unit time or distance.

frequency-domain fluorescence-lifetime imaging microscopy

A mode in fluorescence lifetime imaging microscopy in which a fluorophore is excited with a continuous sinusoidally modulated source, producing fluorescence emission at the same frequency but shifted in phase and reduced in amplitude.

frequency doubling see second harmonic generation

Fresnel contrast

The contrast that arises in images when they are slightly out of focus.

Fresnel diffraction

Diffraction in which the source and the image are a finite distance from the object, so the wavefronts can be treated as curved.

Fresnel fringes

Bright or dark fringes typically observed at the edges of a hole in a defocused specimen, produced by interference of scattered radiation or electrons with the transmitted beam. In an electron microscope the number of fringes is a measure of beam coherence; usually only one fringe is seen in a weakly coherent TEM; up to 100 may be seen in a HREM with a coherent field-emission source. Fresnel fringes produced by holey carbon films are commonly used for astigmatism correction.

Fresnel image

An out-of-focus image in which contrast is derived from Fresnel diffraction.

Fresnel lens

A large converging lens with at least one surface formed by a concentric series of ring-shaped steps, reducing the amount of material needed to form the lens. Fresnel lenses are used in lighthouses and overhead projectors.

friction force microscopy see lateral force microscopy

fringes see interference fringes

front focal length

The focal length on the object side of a lens.

front focal plane

The focal plane on the object side of a lens.

front-illuminated charge-coupled device

A standard configuration for a charge-coupled device, in which the illumination passes through the polysilicon gate structures of a pixel before generating charges.

front lens

The first lens in an objective, nearest the specimen.

frozen hydrated specimen

A specimen that contains water in the form of ice.

frozen section see cryosection

f stop

A synonym for F number used to define the light gathering power of a camera lens with an iris diaphragm. A typical series of preset f stops is: f/1.4, f/2, f/2.8, f/4, f/5.6, f/8, f/11, f/16, f/22; each consecutive stop increases f number by a factor of $\sqrt{2}$, corresponding to a decrease in light flux of one half. Smaller f-stops reduce the light-gathering ability of a lens but increase its depth of field.

full-frame charge-coupled device

A basic type of charge-coupled device with every pixel in the parallel register used for imaging. A full-frame CCD requires a shutter to allow charge read out between exposures.

full-wave plate

A retardation plate that produces an optical path difference or retardation of one wavelength, typically for green light ($\lambda \cong 550$ nm). A full-wave plate produces a red image when placed at 45° between crossed polars illuminated with white light, because green light is retarded 1λ and emerges from the plate polarized in the same orientation as the analyzer and is blocked; other colors experience variable retardation and emerge elliptically polarized and pass through the analyzer. When a retarding object is placed between polarizer and plate, a different wavelength of light is blocked by the analyzer, generating a differently colored image.

full width at half maximum

The width of a peak at one half of its height, a notation used to characterize a single peak or resolve adjacent peaks in a spectrum.

full width at tenth maximum

The width of a peak at one tenth of its height, a notation used to characterize a single peak or resolve adjacent peaks in a spectrum.

functionalized tip see chemical tip

fused quartz see quartz

G

Gabor focus

The value of the defocus for which the transfer of phase information is approximately equal to the transfer of amplitude information.

gain

An amplification or reduction factor applied between the input and output of an imaging device. In analog video, gain is an increase in the gradient of the curve relating video amplifier output to target illuminance.
In a CCD camera, the gain is the standard number of photoelectrons per ADU (analog-to-digital unit); decreasing the gain results in fewer electrons per ADU making the camera more sensitive to low light levels.

galvanometric mirror

A mirror whose motion is controlled by a galvanometer, typically used in optical scanning devices, such as a confocal laser scanning microscope.

gamma

The exponent γ relating the input to output signals of an imaging device such as a video camera or monitor. In a logarithmic plot of the input versus output values, the gamma is the slope of the curve; a gamma of 1 indicates a linear relationship. Camera controls and image processing software allow gamma correction: decreasing gamma highlights low-intensity pixels; increasing gamma highlights high-intensity pixels.

gamma correction see gamma

gamma rays

Electromagnetic radiation with wavelengths in the range 10^{-10} to 10^{-14} m and very high energies in the range 10 keV to 10 MeV. Gamma rays are emitted by excitation of atomic nuclei.

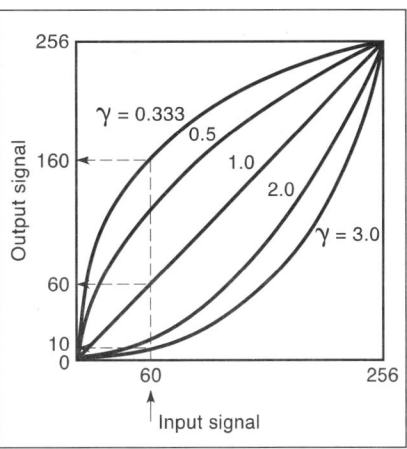

Principle of gamma γ and gamma correction of imaging device. From D. Murphy © John Wiley and Sons, Inc.

gas-discharge lamp

A lamp that produces a light by excitation of gas molecules.

gaseous secondary-electron detector

A secondary electron detector used in an environmental scanning electron microscope. The gaseous secondary-electron detector has an electrode with a positive potential of up to 600 V which attracts secondary electrons emitted by the specimen. As these electrons pass through the column they interact with gas molecules creating additional secondary electrons and positive ions by avalanche amplification of the original signal. This signal is collected by the electrode, processed and read out to a cathode ray tube.

An important feature of the gaseous secondary-electron detector is that the positive ions created in the chamber are repelled by the detector electrode back to the specimen, neutralizing any charging effects.

gas proportional counter

A detector used in wavelength-dispersive spectrometers that measures photoelectrons emitted by a gas. A gas proportional counter comprises a sealed gas-filled metal tube (the cathode) with a thin central wire (the anode); the gas is typically argon. X-rays entering the detector through a thin window ionize the gas, creating photoelectrons that amplify the initial signal by avalanche emission as they are attracted to the anode. The current pulses output by the anode indicate the number of incident X-rays of a specific wavelength.

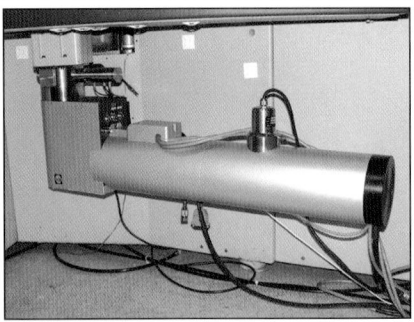

Gatan imaging filter. Schematic showing major components and view of position beneath a TEM. Courtesy of Gatan, Inc.

Gatan imaging filter

A brand-specific post-column energy filter for electron energy loss spectroscopy and electron spectroscopic imaging. The GIF design incorporates computer controlled entrance apertures, entrance optics for focusing and aberration correction, a single 90° sector prism, a computer controlled energy selecting slit, multipole correctors, and a CCD detector.

Gaussian focus

The point or plane of focus of paraxial rays.

Gaussian lens equation see thin-lens equation

Gaussian optics see paraxial optics

gelatin capsule

A cylindrical gelatin mold used for specimen embedding.

Gemini lens

A brand-specific design of electron optics for scanning electron microscopes, featuring a field-emission gun, in-lens secondary electron detection and a magnetic-electrostatic objective lens.

geometrical optics

A method in optics that describes the propagation of light through an optical system without consideration of the phenomena of coherence, diffraction and interference.

germanium detector

A detector made from a high-purity germanium crystal, used in X-ray spectrometers. A germanium detector is typically ~5 mm thick to trap more X-rays, does not require doping, and can be used at high energies (>20 keV). The excitation energy is ~2.98 eV.

Schematic of Gemini lens. Courtesy of Carl Zeiss SMT.

getter pump see ion-getter pump

GIF image file format

The Graphic Interchange Format image file format. File extension: .gif. The GIF format uses 256 colors, can be interlaced and supports transparency making it popular for Web applications.

glass cement see optical cement

glass knife

A knife used to cut thick or thin sections in a microtome or ultramicrotome. Standard triangular glass knives are made from 1/4" thick strips of knife glass using a knifemaker or glass pliers. The strips are broken into squares; each square is then fractured into two knives. The sharpest region of

Triangular glass knives. Courtesy of Boeckeler Instruments, Inc.

the blade lies closest to the conchoidal stress line that runs across the fracture face of the glass square.

gliding stage see **sliding stage**

glow discharge

The creation of a plasma of charged ions, typically used for specimen cleaning and priming before further coating procedures, or to reduce the hydrophobicity of substrates, e.g. support films. Glow discharge can be performed in most coating units at low vacuum, typically ~13 Pa, using a gas such as argon.

glutaraldehyde

1,5-pentanedial, $OHC-[CH_2]_3-CHO$, MW 100.11. A chemical fixative suitable for all biological and some materials specimens. The two aldehyde groups of glutaraldehyde form cross-links between amine groups in lysine-containing proteins and macromolecules.

glutardialdehyde see **glutaraldehyde**

gold autometallography

A mode of autometallography that enhances the contrast of metal probes by the addition of gold atoms.

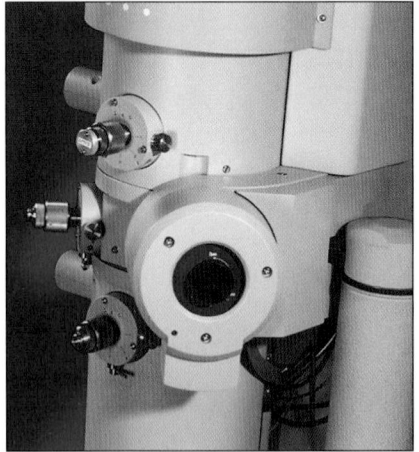

Goniometer on transmission electron microscope. Courtesy of FEI Company.

gold enhancement see **gold autometallography**

gold probes see **colloidal gold, nanogold**

goniometer stage

A standard type of stage on a transmission electron microscope that allows manual or motor-driven axial rotation of the specimen holder. The eucentric plane is adjusted with the goniometer height or Z control.

grain size

The size of the silver halide crystals in the emulsion of photographic film or paper. Grain size affects the speed of the film or paper and the resolution of images.

In standard types of TEM film the grain size is

around 5 μm.
In 35 mm film, grain size is 0.5 to 3 μm,
depending on film speed.

graticule see reticle

grating see diffraction grating

grating equation see diffraction grating

gray level
The intensity of a point in a grayscale image. In a
digital grayscale image the gray level of each pixel
is the numerical value specified by the bits
encoding the pixel. The range of gray levels that
can be stored in a digital image is determined by
the bit depth.

grayscale image
An image having pixels or tones with intensities
varying between black and white.

grazing-exit electron probe microanalysis
X-ray microanalysis using X-rays emitted from the
surface layers of a specimen at very low take-off
angles (~1°).

green fluorescent protein see fluorescent proteins

Greenough stereomicroscope
A type of stereomicroscope that uses separate
optical systems for each eye. The systems contain
erecting prisms and are inclined to produce a
convergence angle of about 10°.

grid box
A box for storage and transport of electron
microscope grids, usually with indexed
compartments.

grid glue
An adhesive used to attach a support film to an
electron microscope grid.

grid see gun grid, support grid

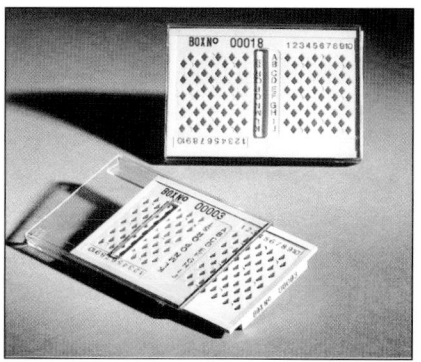

Gridbox. Courtesy of Agar Scientific.

grinder

A device that thins by abrasion. A grinder contains a rotating grinding wheel that thins a specimen by wet grinding using a set of abrasive grits with progressively finer dimensions, down to 0.5 μm. A grinder is typically used to pre-thin specimens before final thinning by dimpling or ion milling.

grinding

The removal of surface material by abrasion.

ground-glass diffuser

A light scrambler formed by roughening the surface of a glass plate by etching or grinding.

ground-glass filter see ground-glass diffuser

gun airlock

The airlock between the gun and column of an electron microscope.

gun bias

The potential difference between the Wehnelt and the filament of an electron gun, typically ~200 V.

gun coils

A set of double deflection coils, located above the condenser lenses, that control the alignment of the beam emerging from the gun of an electron microscope. The controls are usually labeled gun tilt and gun shift.

gun grid

An electrode, made from wire mesh or a perforated plate, that is placed between the anode and cathode of a triode gun to regulate the intensity of the electron beam.

gun lens

A magnetic or electrostatic lens in a field-emission gun that focuses the electron beam.

gun see electron gun

gunshot residue analysis

The use of X-ray microanalysis to investigate gunshot residues.

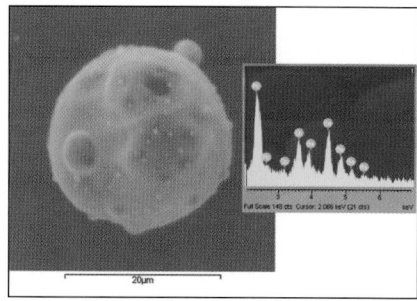

Gunshot residue analysis by energy-dispersive X-ray spectroscopy. SEM image of gunshot particle (left) and X-ray spectrum (right) showing peaks for Ba, Pb and Sb. Courtesy of Oxford Instruments.

H

hairpin filament

A V-shaped filament, such as the tungsten filament used in thermionic electron guns.

halftone

A black and white image created in print media by using a pattern of variably spaced black and white dots to simulate different shades of gray.

half width at half maximum

One half of the width of a peak at one half of its height, a notation used to characterize a single peak or resolve adjacent peaks in a spectrum.

half-wave plate

A retardation plate that produces an optical path difference or retardation of one half wavelength. Linearly polarized light with a plane of vibration at an angle θ to the optic axis emerges rotated by an angle of 2θ.

Hall method

A method for quantifying the elemental content in thin specimens in an analytical electron microscope, typically applied to biological specimens.

In the Hall method concentration C_a is measured as mass of a given element A per unit mass of the specimen. The peak above background intensity of a given element (I_a) measures the mass of that element in the area of analysis, and the counts in the continuum region (W) give a measure of the mass of the specimen:

$$C_a = k\,(I_a/W).$$

The constant k is determined from the analysis of a suitable standard, generally made from a solution of salts in 15-20% protein that has been prepared in a similar manner to the specimen.

halogen lamp see quartz halogen lamp

hard X-rays

X-rays with wavelengths in the range 0.01-1.0 nm, typically used in X-ray diffraction, tomography and radiography.

hardener

A compound added to a resin to increase the hardness after polymerization. The amount of hardener affects the cutting properties of the block. Examples: dodecenyl succinic anhydride and nadic methyl anhydride for Araldite and Epon resins; nonenyl succinic anhydride for Spurr's resin.

HBO lamp see mercury arc lamp

heat filter

A filter that blocks infrared wavelengths, typically used to protect heat-sensitive specimens.

heating stage

A microscope stage that uses resistive heating to maintain specimens at high temperatures.

heavy-metal salts

The salts of lead, molybdenum, osmium, tungsten, uranium and other high atomic number elements used as specimen stains for electron microscopy.

height mode see constant height mode

Heating and cooling stage (range 600°C to -196°C) for light microscope. Courtesy of Agar Scientific.

helium atom microscope

A type of microscope that uses a beam of helium ions. The helium atom microscope is used as a non-destructive tool for surface reflection microscopy.

hertz

SI derived unit of frequency. Symbol Hz. One hertz is one cycle per second.

hexapole see sextupole

hiatus

The distance between the two principal planes of a lens.

high-angle annular darkfield detector

An annular solid-state detector for elastically scattered electrons, used in darkfield imaging mode in scanning transmission and transmission electron microscopes. The detector has a central aperture for the direct beam or a bright field detector and typically detects electrons scattered though semiangles greater than 50 mrad (3°).

high-angle annular darkfield detector

An annular solid-state detector for elastically scattered electrons, used in darkfield imaging mode in scanning transmission and transmission electron microscopes. The detector has a central aperture for the direct beam or a bright field detector and typically detects electrons scattered though semiangles greater than 50 mrad (3°).

high-angle annular darkfield mode

The use of a high-angle annular darkfield detector for the detection of quasi-elastically scattered electrons in a scanning transmission electron microscope. High-angle annular darkfield images typically show mass thickness contrast or atomic number contrast.

high-angle elastic scattering see Rutherford scattering
higher order Laue zones see Laue zones

high-eyepoint eyepiece

An eyepiece with an eyepoint 20-25 mm from the eyelens, designed for use with spectacles.

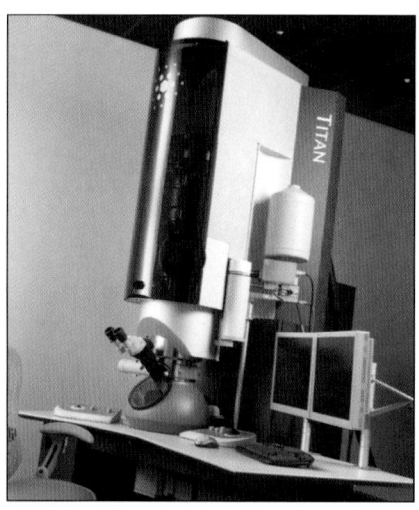

High-resolution transmission electron microscope. Courtesy of FEI Company.

high-loss region

A region in an electron energy-loss spectrum formed by electrons with energy losses of 50 to thousands of eV. The high-loss region contains peaks or ionization edges characteristic of the elements in the specimen, superimposed on a background of energy loss arising from plural scattering of electrons.

high-pass filter

An image filter that reduces low spatial frequency details, sharpening the image.

high-pressure freezing

The cryofixation of a specimen at high pressure. High-pressure freezing reduces the damage to specimens caused by ice crystal formation. At high pressure (~2.1 x 10^8 Pa) the freezing point of water is lower (~-21°C) and the onset of of ice crystal nucleation occurs at lower temperature (~ -91°C) allowing slower cooling rates (~ 200°C s^{-1}).

high-pressure freezing device

A device for the high-pressure cryofixation of specimens. A high-pressure freezing device contains a pressure chamber in which the specimen, sandwiched between a pair of copper plates, is placed. The chamber is first flushed with isopropyl alcohol at high pressure and then rapidly filled with liquid nitrogen at 210,000 Pa.

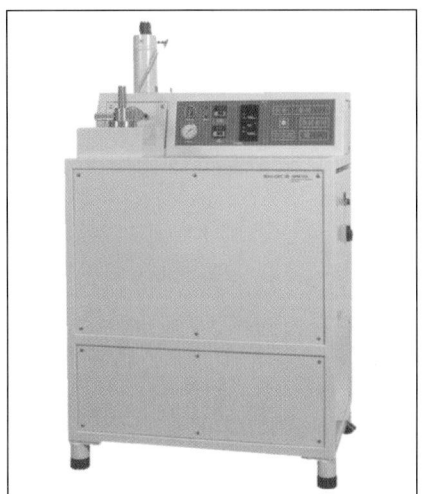

High-pressure freezing device. Courtesy of BAL-TEC AG.

high-purity germanium detector see germanium detector

high-resolution electron microscopy

The use of a transmission electron microscope at the limit of resolution of the instrument, typically to obtain atomic-resolution images and chemical information from electron energy-loss spectroscopy.

high-resolution transmission electron microscopy see high-resolution electron microscopy

high tension

Synonym for high voltage.

high-tension wobbler see high-voltage wobbler

high-tilt holder

A specimen holder designed for electron tomography, with reduced sides and low profile, allowing high (±80°) tilt angles.

high vacuum

A gas pressure in the range 10^{-5} to 10^{-7} pascals.

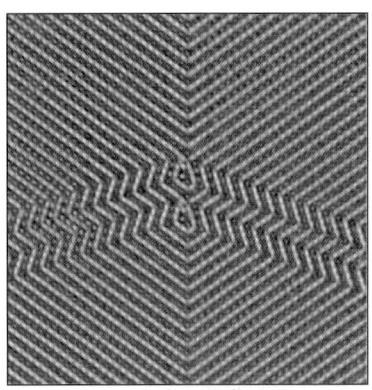

High resolution electron microscopy. Corrected phase image of $BaTiO_3$ grain boundary from through-focus series reconstruction. Courtesy of FEI Company and A. Thust and C. Jia, Forschungszentrum Jeulich, Germany.

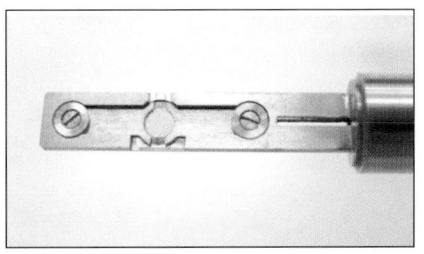

High-tilt specimen holder for electron tomography. Courtesy of Gatan, Inc.

high-voltage electron microscope

A type of transmission electron microscope with high accelerating voltages (1-3 MV). A high voltage electron microscope is used primarily to examine very thick sections of biological and materials specimens. The key benefits of an HVEM are the high beam penetration ($\leq$10 μm), reduced radiation damage, reduced chromatic aberration, and the large volume around the stage for in-situ experiments. A HVEM is a very large microscope, up to 40 feet high, and requires a special facility.

high-voltage supply

The power supply that produces the accelerating voltage for an electron microscope. The stability of operation is critical for high-resolution electron microscopy.

high-voltage wobbler

A control on an electron microscope that alternately increases and decreases (oscillates) the high-voltage supply to allow alignment of the electron beam.

histochemistry

The identification and localization of the chemical and molecular constituents of tissues by stains, probes and indicators.

histogram equalization

A procedure in digital image processing to optimize contrast. In histogram equalization or stretching the input values (or thresholded values) are adjusted to span the full range of output values.

histogram stretching see histogram equalization

histological section

A section ~5-10 μm thick, used in histology.

histology

The study of biological tissues. Histology typically involves the study of tissue sections by light microscopy.

Principle of histogram equalization. (a) Low contrast image and plot of input vs output values. (b) Result of equalization of min. and max. input and output values. From D. Murphy © John Wiley and Sons, Inc.

Hoffmann modulation contrast microscopy

A form of modulation contrast microscopy that turns phase gradients into image contrast generating pseudo-relief images of unstained and birefringent specimens such as living cells in plastic dishes. In a light microscope equipped for Hoffmann modulation contrast microscopy, a diaphragm with an off-axis slit is placed in the front focal plane of the condenser producing oblique illumination of the specimen. A modulator plate with three differently sized neutral-density regions − dark gray (1% transmission), gray (15%) and transparent − is placed in the back focal plane of the objective so that the slit aperture is conjugate with the semitransparent region. Undiffracted zero-order light passes through the gray region of the plate and is amplitude modulated; light diffracted by positive and negative phase gradients passes through the transparent and dark gray regions respectively, accentuating contrast differences on either side of specimen details. Advanced HMC systems use polarized light to enhance contrast.

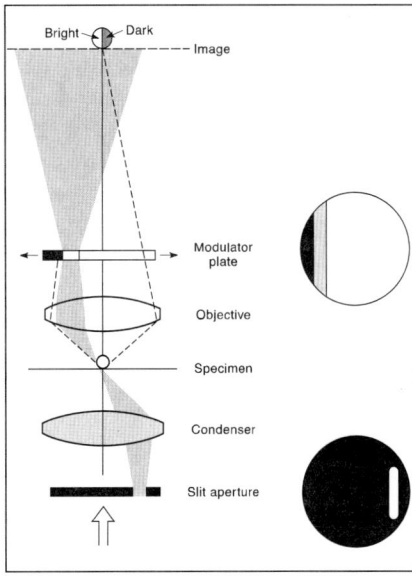

Optical components and pathways in a modulation contrast light microscope. From D. Murphy © John Wiley and Sons, Inc.

hole

A vacant electron position in a solid material that behaves like a positive charge carrier.

hole grid

An electron microscope support grid with a central hole, used for unobstructed viewing of large specimens.

holey carbon film

A carbon film containing holes of variable sizes suitable for examination of specimens by transmission electron microscopy. A holey carbon film permits unobstructed examination of those regions of the specimen overlying the holes (e.g. sections and large specimens) or filling the holes (e.g. particles in a negative stain or vitreous ice). A holey carbon film may also be used for astigmatism correction of transmission electron microscopes.

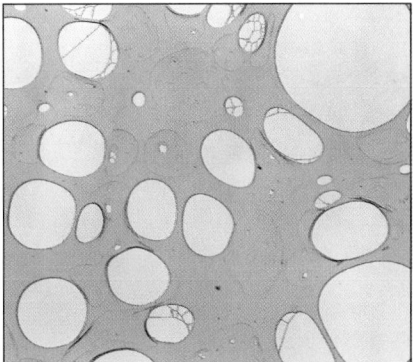

Holey carbon film. Courtesy of Agar Scientific.

hologram

An image that contains both amplitude and phase information from a specimen. A hologram is formed by interference of two coherent beams: a reference beam (e.g. a laser or electron gun) and an object beam (often from the same source) that has been diffracted by the specimen. Illuminating a transparency of the hologram from the rear with a coherent beam reconstructs the object beam, forming twin images - a virtual image on the source side and a real image on the viewer's side - allowing an observer looking through the hologram to see the specimen as a three-dimensional image.

holography

The production of a hologram.

holomicrography

The production of a hologram using a light microscope.

horizontal field width

The distance in an object that corresponds to the full width of a displayed or printed image. Horizontal field width is an alternative to a scale bar in published micrographs.

hot-cathode gauge

An ionization pressure gauge that uses thermionic filament to generate electrons.

hot pixels see warm pixels

HSL system

A system for defining colors based on their hue, saturation and luminosity (lightness).

hue

The attribute of color, as measured by wavelength.

Huygenian eyepiece

An internal-diaphragm light microscope eyepiece with two planoconvex lenses, widely used in laboratory light microscopes.

Huygens' construction see Huygens' principle

Huygens' principle

Huygens' principle states that every point on a
wavefront can be regarded as a source of
secondary spherical waves (wavelets) of equal
frequency and speed. A tangent linking any set of
wavelets defines the position of the wavefront at
any subsequent point and time.

Huygens' wavelets see **Huygens principle**

hypermetropia

An aberration of the eye in which parallel rays
from distant objects are focused behind the retina;
the near point is further away than normal.
Hypermetropia can be corrected with a positive
spectacle lens.

hyperopia see **hypermetropia**

hypo see **fix**

hysteresis

The dependence of the relationship between two
physical properties on whether one property is
increasing or decreasing. Hysteresis occurs in
magnetic lenses and can result in a difference in
the magnification at a given lens setting arrived at
by increasing or decreasing lens excitation.

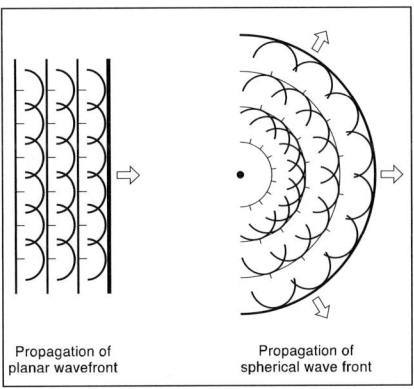

Propagation of planar wavefront

Propagation of spherical wave front

Use of Huygens's principle to describe
propagation of wavefronts. From D.
Murphy © John Wiley and Sons, Inc.

I

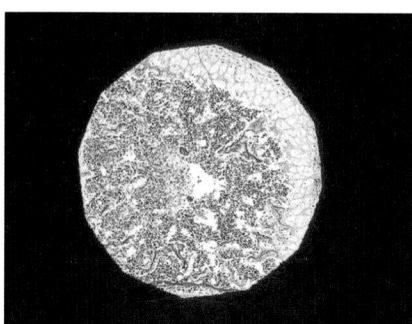

Specimen and partially closed illuminated field iris diaphragm. Courtesy of Carl Zeiss Ltd.

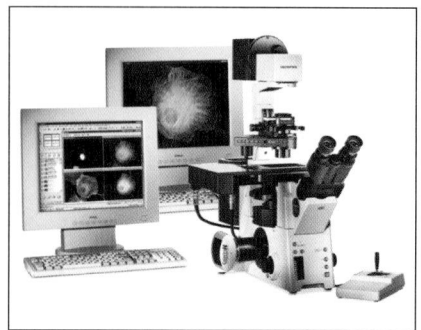

Image analysis system. Courtesy of Soft Imaging System.

Iceland spar see **calcite**

illuminance
A photometric quantity used to characterize the luminous flux per unit area. Symbol E. Units: lumens per square meter ($lm\ m^{-2}$), or lux.

illuminated field diaphragm
A variable diaphragm that regulates the area of the specimen that is illuminated in a light microscope.

illumination
The radiation incident to an object.

image
A representation of an object derived from its interaction with radiation.

image acquisition
The capture of an image by a detector, e.g. photographic film or charge coupled device.

image analysis
The use of mathematical procedures to extract information of any kind from an image.

image-analysis system
A system comprising a microscope, camera, computer, display monitor and image analysis software.

image archiving
The storage of images, typically on microfiche or electronic media such disks and tape.

image archiving software
Software for the management of a computer image database, including tools for indexing, annotation and retrieval.

image circle
The diameter of the image formed by a circular lens at a detector.

image coils

A set of double deflection coils, located beneath the objective lens, that control the position of the image on the screen or detector. Controls are usually labeled image tilt and image shift.

image compression

The reduction of the number of bits used to encode a digital image without reducing its dimensions or unacceptably degrading the image. Lossless compression reduces file sizes but keeps all the original information. Lossy compression reduces file sizes but removes some original information.

Image archiving software. Courtesy of Soft Imaging System.

image coupling see image mode

image database

An image archive, typically stored in digital form on a computer.

image distance

The distance on the optical axis between the image plane and the image-side principal plane.

image enhancement

The use of image processing techniques that improve the visual appearance of an image without using any detailed information about the image-forming process. Image enhancement methods include the use of image kernels to sharpen edges, reduce the effect of noise and match the range of grey levels to the human visual response. Some methods use linear operators (convolutional techniques), others use the nonlinear operators of mathematical morphology.

image filter

A mathematical operation performed on a digital image that modifies the values of pixels in a digital image in order to enhance details such as contrast and edges or remove details such as noise.

image intensifier

An electronic device that enhances the luminance of an image. An image intensifier is usually placed in front of imaging detector, e.g. a microchannel plate image intensifier for a CCD.

Proximity focused image intensifiers with GaAsP photocathodes. Courtesy of Hamamatsu.

Image montaging: The six separate images of a semiconductor in the background were stitched together to form the single image in the foreground. Courtesy of Soft Imaging System.

image interference microscopy

A mode in widefield fluorescence microscopy in which the specimen is imaged from opposite sides through two high-NA microscope objectives focused on the same specimen plane; the two images are then combined at the detector generating an interference pattern containing high resolution axial information.

image kernel see kernel

image mode, in EELS

The use of a diffraction pattern of a specimen as the object for the lens of an energy-loss spectrometer; a standard mode for energy-loss electron spectroscopy.

image montaging

The formation of one large image from many smaller ones. Image montaging is typically used when a specimen is larger than the field of view of a microscope.

image plane

The plane that contains the image formed by an optical system.

image plate

An image recording device comprising a plastic sheet coated with a thin film of crystals (e.g. barium fluoro-bromide) that stores an image after exposure for several days. Upon exposure the crystals are excited to a non-radiative photostimulatable state; when illuminated by a red laser beam the crystals become luminescent and emit blue light in proportion to the original exposure. Imaging plates are reusable.

image-plate detector

A detector with an image plate and laser scanner to read out the images.

image point

The point on the optical axis intersected by the image plane.

image processing

A large family of mathematical operations performed on images in order to improve them in some way, to identify and measure certain features or to extract latent information. Image processing is conventionally divided into four broad areas: image acquisition, sampling and coding; enhancement; restoration; and image analysis and pattern recognition.

image restoration

The use of digital image-processing techniques that exploit all available information about the image-forming process to produce as near as possible a 'true' image of a specimen, free from any degradation caused by contributions from out-of-focus objects, microscope aberrations or detector noise.

image rotation, in electron microscopy

The rotation of an image about the optical axis of a electron microscope. Image rotation occurs because electrons follow a helical path through magnetic lenses, so images will rotate (or even invert) when lens excitation is changed, e.g. during focusing or change of magnification. Some EM manufacturers add an extra lens to correct for image rotation when changing magnification.

image-rotation center

The axis of image rotation in an electron microscope.

image simulation

The generation of an image of a model specimen. Image simulation techniques are used in high-resolution electron microscopy and other imaging techniques as an aid to the interpretation of experimental images.

image space

The space on that side of a lens where the image is formed.

image stitching see image montaging
image wobbler see focus wobbler
imaging filter see Gatan imaging filter

Immunoelectron microscopy. 10-nm diameter immunogold particles identifying lactase on microvilli of enterocyte. Courtesy of Julian Heath.

immersion freezing see **plunge freezing**

immersion lens

An electrostatic lens having an overall accelerating or decelerating effect (unlike an einzel lens, which has no such effect). The term immersion lens is commonly used to describe the accelerating field at the cathode [or specimen-emitter] of an emission microscope.

immersion lens see **oil-immersion objective, oil-immersion condenser**

immersion medium

The liquid placed between an immersion objective and the specimen or coverslip, e.g. oil, water.

immersion oil

A low-viscosity oil with a refractive index n of 1.515, matching that of standard coverslips. Specialized immersion oil ($n = 1.78$) is used with total internal reflection fluorescence microscopy objectives.

immersion refractometry

Refractometry by refractive index matching. The unknown is placed in a series of media of differing refractive indices; a match is made when the object has minimum contrast.

immunocytochemistry

The identification and localization of the molecular constituents of cells using antibodies as probes.

immunoelectron microscopy

The use of electron microscopy for the identification and localization of antigens using immunological probes.

immunoenzyme cytochemistry

The identification and localization of the molecular constituents of cells using antibody and enzyme probes.

immunofluorescence

The fluorescence of specimens labeled with antibody probes conjugated to fluorophores.

immunogold
Colloidal-gold particles conjugated to antibody probes.

immunogold-silver staining
A form of autometallography for the enhancement of ligand-conjugated gold particles with silver.

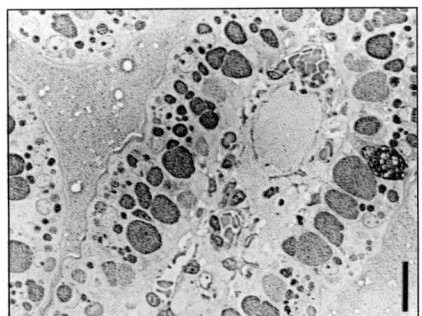

Immunogold-silver staining. Silver-enhanced goldlabeled immunocytochemical localization of pig IgG. Courtesy of Julian Heath and Laszlo Komuves. Scale bar = 50 μm.

immunohistochemistry
The identification and localization of the molecular constituents of tissues using antibodies as probes.

immunohistology
The identification and localization of the molecular constituents of tissues using antibodies as probes.

immunolabeling
The identification and localization of the molecular constituents of a specimen using antibodies.

impulse response
A measure of how much any device degrades signals between reception and output. The instrument impulse response of a diffraction-limited microscope to a point object is the point-spread function.

incandescent lamp
A lamp with a tungsten filament. An incandescent lamp emits a continuous spectrum of visible light with high emission of infrared.

included angle
The angle between the front and rear surfaces of the blade of a microtome knife.

incoherent interference illumination
The illumination of a fluorescent specimen from opposite sides through two high-NA microscope objectives generating an axial interference pattern of excitation light similar to that in standing wave microscopy.

incoherent interference illumination image interference microscopy

A mode of widefield fluorescence microscopy that combines incoherent interference illumination with image interference microscopy. A focal series of images is collected and deconvolved to produce a 3D image with high (<100 nm) axial resolution.

incoherent scattering

The loss of coherence of waves and electrons due to high-angle elastic scattering and inelastic scattering.

in-column camera

A film or CCD camera that enters through the side of the column of a transmission electron microscope.

in-column energy filter

An electron spectrometer situated in the column of a transmission electron microscope.

incomplete charge collection

The failure of a semiconductor detector to release all charge carriers for measurement.

index of refraction see refractive index

indirect labeling

The use of another probe, or set of probes, to locate a primary probe in cytochemistry. Indirect labeling is commonly used to amplify signals.

inductively coupled plasma

A plasma produced by electrical currents induced by fluctuating magnetic fields.

inelastic cross-section see inelastic scattering cross-section

inelastic scattering

The scattering of radiation with loss of energy. Inelastic scattering occurs as a result of interactions with atomic shells. The lost energy generates heat or is used in excitation processes.

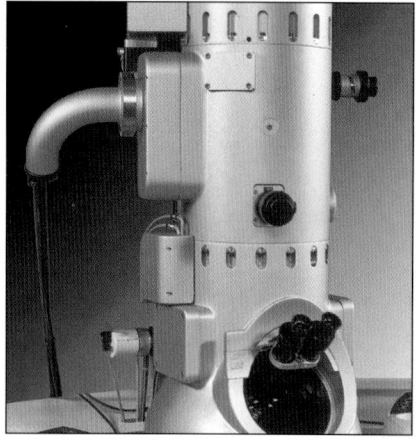

In-column energy filter on a transmission electron microscope. Courtesy of Carl Zeiss SMT.

inelastic scattering cross-section

The cross-section of an atom that yields an inelastic scattering event.

infiltration

The permeation of an embedding medium into a specimen.

infinity-corrected objective

An objective designed for use in a light microscope with infinity-corrected optics.

infinity-corrected optics

A design for the optical pathway of modern light microscopes which allows the introduction of multiple optical accessories between the objective and eyepieces without change of magnification. In infinity-corrected optics, light emerges from the objective as parallel, not converging, rays focused at infinity. After passing through any accessories such as zoom magnifiers and filter cubes, the rays are collected by a tube lens which forms the real intermediate image at the eyepieces.

infinity optics

An ambiguous abbreviation for infinity corrected optics.

infinity viewing

A technique for the unaided viewing of parallel stereo pairs of images. In infinity or free viewing the eyes are relaxed as if gazing into the distance; a stereo pair placed in front of the eyes can then (with practice) appear as three images, with a central stereoscopic image.

infrared

Electromagnetic radiation with wavelengths in the range 700 nm to 1 mm.

infrared absorption

The absorption of infrared radiation due to vibrational and rotational transitions within molecules, typically changes in dipole moments of ionic bonds.

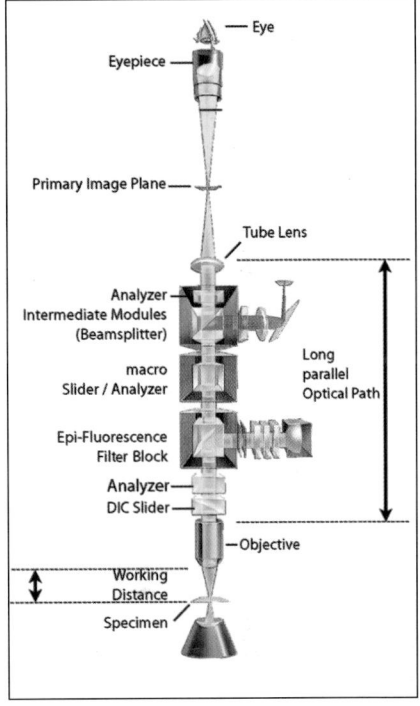

Optical configuration of a light microscope with infinity-corrected optics. Courtesy of Nikon.

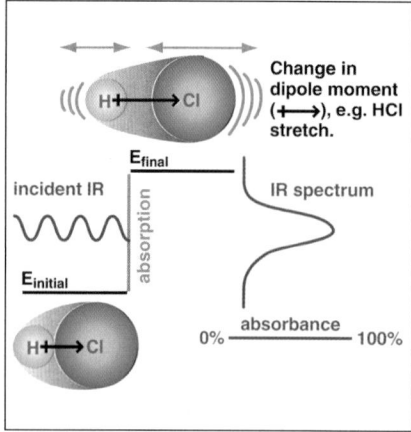

Principle of infrared absorption and infrared spectroscopy. Courtesy of Renishaw plc.

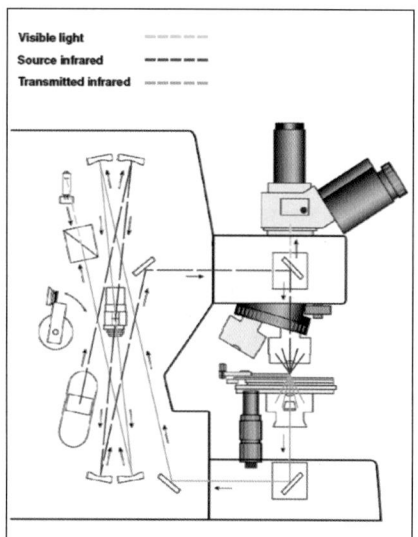

Optical pathways in a transmitted infrared microscope. Courtesy of Thermo Electron Corporation.

infrared absorption spectrum

A plot of the intensity of emitted infrared radiation at each wavelength or wave number. In infrared spectroscopy, the absorption bands in the spectrum are compared to a reference spectral database to identify specific molecular bonds that identify the chemical composition of the specimen. The intensity of each band is proportional to the number of absorbing molecules.

infrared filter see heat filter

infrared map

An image in which each pixel is color coded by degree of infrared absorption.

infrared microscope

A microscope that uses infrared radiation to form an image and investigate specimens. An infrared microscope is typically used to characterise specimen structure by dispersive or Fourier-transform infrared spectroscopy, or to examine specimens that are sensitive to visible light.

infrared microscopy

The use of an infrared microscope to image or investigate a specimen.

infrared objective

A light microscope objective designed for infrared spectroscopy. The key features of an IR objective are lenses with high transmission in the infrared, and additional optics and attachments for specialized techniques such as ATR-IR.

infrared reflectance microscopy

Infrared spectroscopy using radiation that is reflected by the specimen.

infrared spectrometer see Fourier-transform infrared microscope

infrared spectroscopy

The characterization and quantification of molecules by absorption of infrared radiation.

infrared transmission microscopy

Infrared spectroscopy using radiation that is transmitted through the specimen.

in-lens detector

A secondary or backscattered electron detector that is situated inside the condenser/objective lens of a scanning electron microscope.

in-line holography

The generation of a hologram by interference of co-axial reference and object beams after they have both passed through the specimen.

inner-shell ionization edge see ionization edge

in-situ electron microscopy

The experimentation on specimens while inside an electron microscope, such as the use of heating or straining stages.

in-situ hybridization

The identification and localization of nucleic acids (genes, DNA and messenger RNA) by hybridization with specific molecular probes.

instrument impulse response see impulse response

intensified video tube

A video tube with an image intensifier placed between the primary image and the target.

intensifier silicon-intensifier target camera see silicon-intensifier target camera

intensity

1. The strength of radiation.
2. The value of a parameter.

interband/intraband transition peaks

Peaks in the low-loss region of an electron energy-loss spectrum contributed by excitation and inter- and intraband transitions of valence electrons.

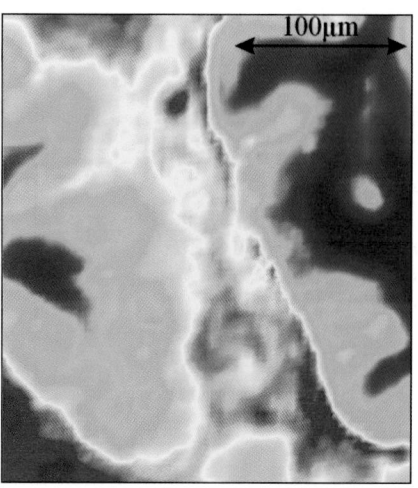

Infrared map of protein amide I band in carcinoma section. *Microscopy and Analysis* 2002.

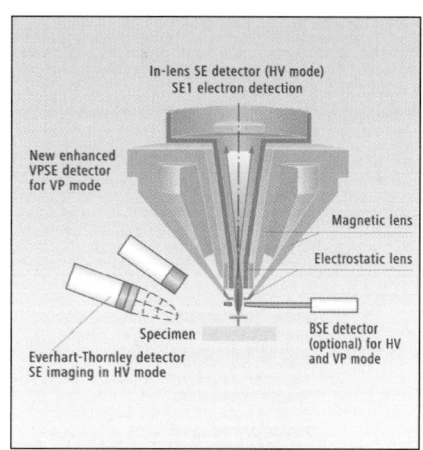

Principle of in-lens SE and BSE detection in a scanning electron microscope. Courtesy of Carl Zeiss SMT.

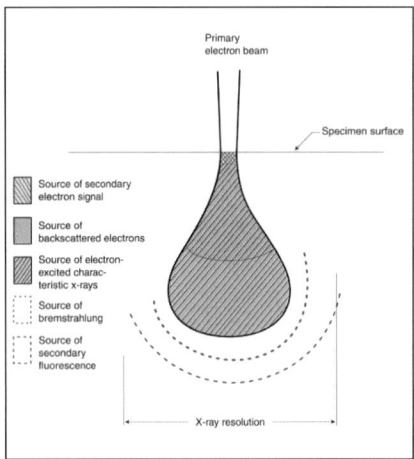

Interaction volume of bulk specimen.
Courtesy of Thermo Electron Corporation.

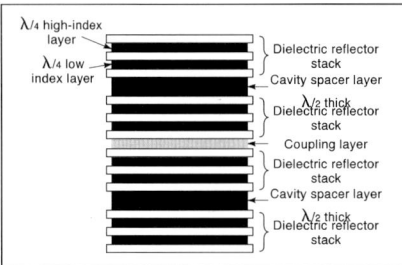

Principle and design of interference filters.
From D. Murphy © John Wiley and Sons,
Inc.

interaction cross-section see scattering cross-section

interaction volume

The volume of an object that interacts with an electron probe. In a bulk specimen the interaction volume is shaped like a teardrop, several micrometers deep, but its actual dimensions are a function of probe energy and specimen composition, generally increasing with higher accelerating voltage (kV). In a thin specimen, the interaction volume decreases with higher kV. The interaction volume determines spatial resolution in X-ray microanalysis.

interference

The superposition of two or more waves of similar frequency or wavelength. Interfering waves produce a resultant wave whose amplitude is the sum of the interacting waves. Interference is the key phenomenon underpinning image formation in microscopy and optical filter design.

interference colors

Colors that are generated by the removal of selected wavelengths from white light by destructive interference.

interference colors, of thin sections

Thin sections of resin-embedded specimens appear colored when floating on a liquid due to the interference colors generated by white light reflected from their upper and lower surfaces. These interference colors may be used to estimate section thickness; grey: ~60 nm; silver: ~ 90 nm; gold: ~120 nm; purple ~200 nm.

interference filter

A filter that uses constructive and destructive interference to regulate the wavelengths of transmitted light. Interference filters contain many reflective metallic films separated by layers of dielectrics with thicknesses either $\lambda/2$ or $\lambda/4$ of the wavelength λ to be transmitted. Light entering the filter is multiply reflected between each layer but only the specified wavelength will progress, by constructive interference; shorter or longer wavelengths are attenuated by destructive interference.

interference fringes

Alternating bright and dark zones in an image produced by constructive and destructive interference, respectively.

interference microscope

A type of light microscope that uses dual-beam interferometry to measure optical pathlength differences between an object beam that interacts with the specimen and a reference beam. The term interference microscope is usually reserved for those microscopes that can measure the optical pathlength differences of optically isotropic specimens. The two common configurations of interference microscopes are: 1. the transmitted-light interference microscope, which is typically used to measure optical pathlength differences of light passing through biological materials such as cells and tissues; and 2. the reflected-light interference microscope which measures surface topography and roughness of opaque specimens.

interference microscopy

Any form of microscopy in which image contrast is manipulated by specialized optical components that regulate interference.

interference objective

A light microscope objective that contains an integral interferometer to measure optical path differences.

interference reflection microscopy

A mode of interference microscopy used to measure the separation of a specimen and a substrate by interference of the light reflected from the apposed surfaces. Interference reflection microscopy is used to study cell-substratum adhesions of living cells on glass coverslips. The cells are epi-illuminated with polarized monochromatic light; rays reflected from the cell-aqueous medium and medium-glass surfaces interfere and generate an amplitude contrast image. Because there is a $\lambda/2$ phase shift in light reflected from the cell-medium interface but no phase shift at the medium-glass interface, destructive

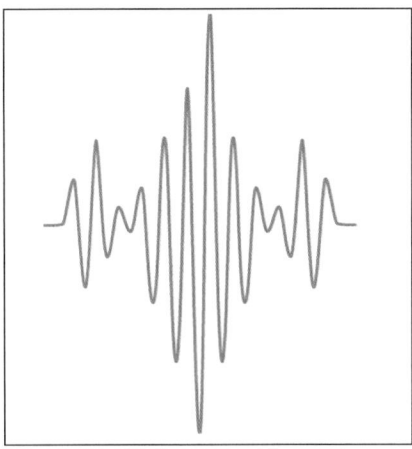

Interferogram. Courtesy of Thermo Electron Corporation.

interference will occur when the surfaces are in contact; constructive interference occurs when the separation gap is $\lambda/4$, i.e. about 120 nm.

interferogram

The signal output from an interferometer, produced by interference of the primary and the reference beams and containing information (the Fourier transform) on the phase and amplitude of the spectrum of the input radiation.

interferometer

A term for a family of instruments that use interference to measure or exploit the optical pathlength differences between an object beam and a reference beam.

interferometric microscope

An interference microscope capable of measuring optical pathlength differences between two spatially separated paths through an optically isotropic specimen.

interferometry

The study and applications of the phenomenon of interference of waves. Interferometry is typically used to measure optical pathlength differences and to derive quantitative data on the properties of matter.

interlaced video

A video signal that forms a full frame by combining two or more video fields.

interline-transfer charge-coupled device

A type of CCD with a parallel register of alternating rows of imaging pixels and masked pixels for storage and read out. In operation, the imaging rows acquire charges which are then rapidly transferred to the adjacent storage rows; charges are read out while the imaging rows are accumulating new charges. Interline-transfer CCDs operate without a shutter and produce a video-rate signal; some are equipped with microlenses to focus light onto the imaging pixels.

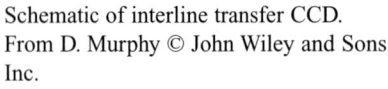

Schematic of interline transfer CCD. From D. Murphy © John Wiley and Sons, Inc.

intermediate contact mode see tapping mode

intermediate image

An image lying in an image or object plane of a multi-component optical system. In a light microscope the intermediate image is the object for the eyepieces. In a transmission electron microscope there may be several intermediate images formed by the objective, intermediate and projector lenses.

intermediate lens

A lens in an optical system that has as its object the primary image and projects an image towards another lens, e.g. the intermediate lens of an electron microscope.

intermediate tube

That part of the tube of a light microscope placed between the body tube and the viewing head that contains additional optical components such as analyzers and filter cubes.

intermediate-voltage electron microscope

A transmission electron microscope with an accelerating voltage in the range 300 to 400 kV, commonly used for imaging of thick sections of biological and materials specimens. The intermediate-voltage electron microscope offers some of the benefits of a high-voltage electron microscope (greater penetration, reduced specimen damage, reduced chromatic aberration) since the improvement in instrument performance between 400 and 1000 kV is slight.

intermediate-voltage electron microscopy

The investigation of a specimen with an intermediate-voltage electron microscope.

intermittent contact mode see tapping mode

internal field-diaphragm eyepiece

An eyepiece with a diaphragm between the field lens and the eyelens.

internal reflection

The reflection of incident light when passing from a medium of high optical density (refractive index) to one of lower density.

internal transmittance see **transmittance**

interocular distance see **interpupillary distance**

interpupillary distance

The distance between the centers of the pupils of the eyes when viewing at infinity; the average distance is about 62 mm. The interpupillary distance of binocular eyepieces can be changed from 55 to 75 mm to suit different observers.

intersection distances

The distances on the optical axis between the object or image planes and their respective lens vertices.

intrinsic germanium detector see **germanium detector**
intrinsic region see **semiconductor detector**

inverse fluorescence recovery after photobleaching

A variant of fluorescence recovery after photobleaching in which fluorophores surrounding a small fluorescent region are bleached and the loss of fluorescence from the unbleached region is analyzed.

inverted light microscope

A light microscope in which the specimen is observed from beneath the stage. An inverted light microscope has the light source and condenser above the stage and the objectives and associated imaging components underneath. This design is particularly suited for experimental treatment of the specimen during microscopy.

ion

An atom or molecule that has gained or lost one or more electrons. A positively charged ion is a cation; a negatively charged ion is an anion.

ion beam

A beam of ions, i.e. atoms that have gained or lost electrons. Beams of positively charged ions, such

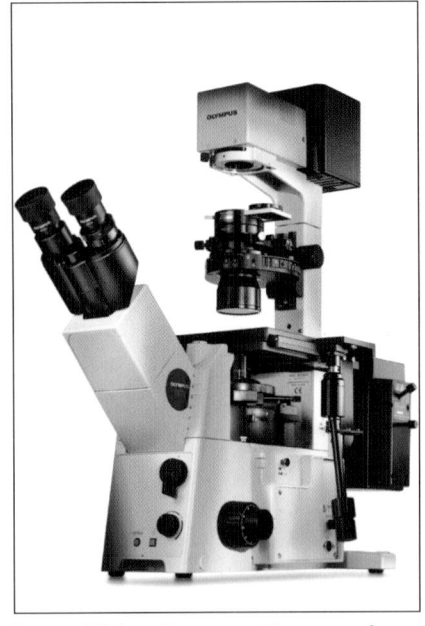

Inverted light microscope. Courtesy of Olympus.

as Ar^+ or Ga^+, are used in ion milling devices and focused ion beam systems.

ion detector

A detector of secondary ions, typically used in secondary-ion mass spectrometry.

ion image detector see **ion detector**
ion microscope see **field-ion microscope**
ion milling see **ion-beam thinning**
ion polishing see **ion-beam milling**
ion-beam milling see **ion-beam thinning**

ion-beam sputter coating unit

A sputter coater that uses an ion beam to sputter the target, producing an ultrafine grain coating suitable for high resolution electron microscopy.

ion-beam sputtering

Sputtering with an ion beam.

ion-beam thinning

The sputtering of a surface with an ion beam.

ion-beam thinning device

A device for the thinning of materials by sputtering with an ion beam. In an ion-beam thinning device the specimen is placed in a vacuum chamber on a liquid-nitrogen cooled stage and thinned by sputtering from both sides with argon ion beams. Typical thinning rates are several micrometers per hour.

ion-getter pump

A vacuum pump that uses chemiadsorption to trap gases using a magnetically confined plasma discharge. Gases are ionized and attracted to a titanium cathode where they are buried, reducing gas pressure. During this process titanium atoms are sputtered and getter additional gas molecules before condensing on pump surfaces. Ion pumps normally operate when the vacuum less than 10^{-4} Pa and can reduce the vacuum to 10^{-9} Pa.

ionization

The ejection of an electron from an inner atomic shell to the vacuum level (in an isolated atom) or to an unfilled state above the Fermi level in a solid.

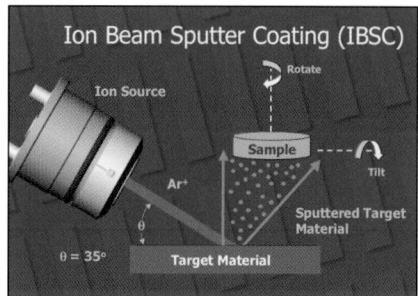

Principle of ion-beam sputtering. Courtesy of E.A. Fischione Instruments, Inc.

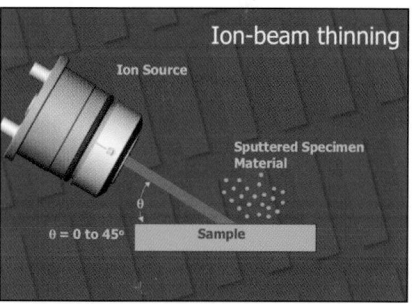

Principle of ion-beam thinning. Courtesy of E.A. Fischione Instruments, Inc.

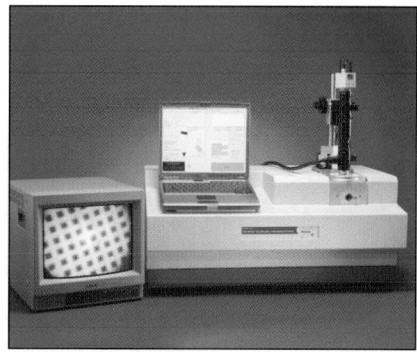

Ion-beam thinning device. Courtesy of E.A. Fischione Instruments, Inc.

183

ionization cross-section

The cross-section that an atom presents for ionization.

ionization edge

A peak in an electron energy-loss spectrum representing a specific ionization event in an inner atomic shell. Ionization edges identify specific elements.

ionization energy

The energy required to ionize an atom.

ionization gauge

A pressure gauge that measures the current generated by the ionization of gases using a hot or cold cathode.

iris

The variable diaphragm of the eye. The iris contains radial and circular muscles that regulate the size of the pupil.

irradiance

A radiometric quantity used to characterize the intensity of electromagnetic radiation. Symbol E; units: watt per square meter.

ISO film rating

The International Standards Organization standard arithmetical scale for the speed or sensitivity of black and white and color film. The higher the number the more sensitive (faster) the film. The ISO film rating has now superseded the similar ASA and the logarithmic DIN ratings.

isoplanatic

A definition of an optical system free of coma and other aberrations except spherical aberration.

J

Jablonski diagram

A graphical representation of molecular electronic and vibrational energy states and their transitions during luminescence.

Jamin-Lebedev interference microscope

A dual-beam interference microscope in which the object and reference beams are orthogonally polarized.

jet electropolishing

Electropolishing with a jet of electrolyte.

jet electropolishing device

An apparatus for the electrolytic thinning and polishing of conductive specimens. The key features of a jet electropolisher are: a cooled electrolyte chamber; a recirculating pump; two electrolyte jets; and a light source and detector. The specimen is mounted between two nozzles that apply jets of electrolyte to each side of the specimen. The rate of thinning is monitored by the change in transmission of a light beam through the specimen.

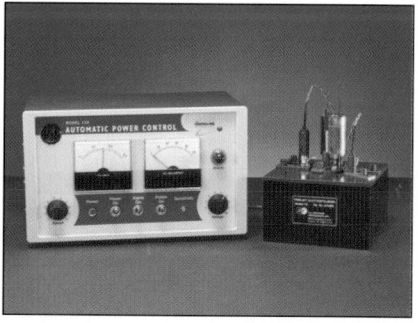

Jet electropolishing device. Courtesy of E.A. Fischione Instruments, Inc.

Johann geometry

A design for the diffracting crystal in a crystal spectrometer. In the Johann geometry the crystal is bent with a radius $2r$ where r is radius of the Rowland circle.

Johansson geometry

A design for the diffracting crystal in a crystal spectrometer. In the Johansson geometry the crystal is bent to a radius $2r$ and its front surface is ground to a radius $1r$, where r is the radius of the Rowland circle.

JPEG image file format

The Joint Photographic Experts Group image file format. File extensions: .jpg, .jpeg, .jpe. The JPEG file format uses adjustable lossy compression.

K

Karnovsky's fixative

A chemical fixative containing formaldehyde and glutaraldehyde. The fixative was originally formulated with 4% FA and 5% GA but is now commonly used with lower concentrations. Karnovsky's fixative combines the advantages of the rapid penetration of formaldehyde with the efficient crosslinking of glutaraldehyde.

karyotyping

The characterization of the number, size, shape of chromosomes by microscopical techniques.

Kasha's rule

Kasha's rule states that excited molecules in solids and liquids relax to the S1 energy state before emitting a photon. A consequence of Kasha's rule is that the emission spectrum of a fluorophore is always the same regardless of the excitation wavelength.

Kellner eyepiece

An external diaphragm light microscope eyepiece with an achromatic doublet as the eyelens.

kelvin

SI base unit of temperature. Symbol K (not °K). A kelvin is equal to one 273.16th of the thermodynamic temperature of the triple point of water (~0°C). 0K is absolute zero.

Kelvin probe microscopy see scanning surface potential microscopy

Kerr Effect

The development of double refraction when an optically isotropic material in placed in a electric field.

kernel filter see filter kernel
k factor see Cliff Lorimer ratio

Kikuchi band see **Kikuchi lines**

Kikuchi diffraction
The diffraction of electrons following diffuse inelastic electron scattering.

Kikuchi lines
Pairs of bright and dark lines in an electron diffraction pattern produced by Kikuchi diffraction. The brighter line is called the excess line, produced by electrons closer to the forward direction, and the darker line is the deficit line; the lines and the intervening space are called a Kikuchi band.

Kikuchi lines in EBSD diffraction pattern. Courtesy of Oxford Instruments.

Kikuchi pattern see **Kichuchi lines**
kiloelectronvolt see **electronvolt**

kinematic diffraction
Electron or X-ray diffraction of very thin specimens where there is no double or multiple diffraction.

knife glass
High quality glass shaped by rolling, so that any stresses are axial, and used for making glass knives for (ultra)microtomy.

knife holder
A component of a microtome or ultramicrotome that holds the knife. The knife holder allows adjustment of the clearance angle and orientation of the knife edge to the block face.

knife marks
A sectioning artifact caused by defects of knife or block leading to section scratches or tears perpendicular to the knife edge. Causes of knife marks include a dull or chipped knife edge, poor specimen infiltration, and dense objects in the specimen.

Ultramicrotome knife holder with glass knife. Courtesy of Boeckeler Instruments Inc.

knife pliers
A pair of glazier's pliers used to break glass strips and squares to form knives for microtomy.

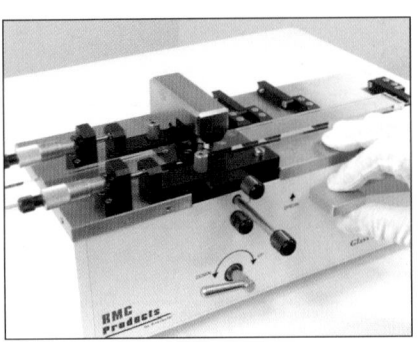

Glass knifemaker. Courtesy of Boeckeler
Instruments Inc.

knife see **microtome knife**
knife trough see **water trough**

knifemaker

A device for the manufacture of glass knives for
microtomy. Knifemakers generally use the
balanced break method to produce exact squares
from a strip of knife glass; each square is then
fractured into two glass knives using the free break
method.

knock-on damage

The displacement of atoms from a crystal lattice
by incident radiation.

Köhler epi-illumination

The use of Köhler illumination with an epi-
illumination or reflected light microscope.
objective. In Köhler epi-illumination a lens in the
epi-illuminator projects an image of an aperture
diaphragm into the back focal plane of the
objective, which acts as a condenser, and the
illumination field diaphragm is in a plane
conjugate with that of the collector lens.

Köhler illumination, in light microscope

An optical configuration that provides uniform
specimen illumination in a light microscope. In
Köhler illumination a collector lens projects an
image of the light source into the aperture
diaphragm in the front focal plane of the
condenser, and the condenser projects an image of
the illuminated field diaphragm into the specimen
plane. This arrangement allows independent
operation of the field and aperture diaphragms to
control resolution and contrast.

Köhler illumination, in TEM

A mode of specimen illumination in a transmission
electron microscope, analogous to Köhler
illumination in light microscopy, allowing
specimen illuminated with a parallel beam of
electrons and independent adjustment of aperture
angle and illuminated area.

Kossel cones

The cones formed by Kikuchi-diffracted electrons
that give rise to Kikuchi lines. Since primary

scattered electrons travel in random directions
from the scattering point, only those electrons that
satisfy Bragg's law will be diffracted, forming two
cones on either side of a crystal plane.

Kossel-Möllenstedt pattern see
nanodiffraction pattern

Kossel pattern

An electron diffraction pattern produced by
convergent beam electron diffraction comprising a
set of overlapping diffraction disks.

Kramers' constant

A constant K used in X-ray microanalysis that
relates the number N of bremsstrahlung X-ray
photons of energy E to the average atomic number
Z of the specimen and the energy E_o of the
incident electron beam in the expression:

$$N(E) = KZ(E_o-E)/E.$$

L

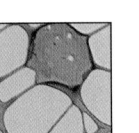

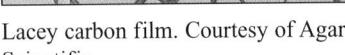

Lacey carbon film. Courtesy of Agar Scientific.

laboratory light microscope

A light microscope with the minimum specifications, suitable for routine applications and training.

lacey carbon film

A carbon support film with large irregularly shaped holes suitable for examination of particulate specimens by transmission electron microscopy.

lamp housing

The housing of a light source on a light microscope. The housing typically contains a socket for the bulb, a concave mirror behind the bulb, a focusable collector lens in front of the bulb, controls for alignment, and vents to allow convective cooling.

lamphouse see lamp housing

lanthanum hexaboride filament

A filament used in the thermionic gun of an electron microscope, consisting of a sharpened crystal of LaB_6 with a 15-μm wide flat tip vertically mounted on a standard filament base.

Lanathanum hexaboride filament. Courtesy of Agar Scientific.

large-angle convergent beam electron diffraction

A mode of convergent beam electron diffraction in which the specimen is moved away from the plane conjugate with the recording plane, permitting information from both the image and the diffraction pattern to be observed and recorded at the same time. LACBED reveals direct and reciprocal space information simultaneously and is particularly useful for the detailed characterization of crystal defects.

large-angle elastic scattering see Rutherford scattering

laser

A device that emits radiation by light amplification by stimulated emission of radiation. In a laser, atoms are excited by pumping and then by interaction with a photon stimulated to emit a photon of similar frequency, phase and polarization, a process called stimulated emission. Lasers produce intense, collimated, coherent, monochromatic beams of light. Types of lasers include gas, solid and semiconductor.

laser beam-induced current see optical beam-induced current
laser diode see diode laser
laser scanning confocal microscope see confocal laser scanning microscope
laser scanning microscope see scanning laser microscope

lateral chromatic aberration

The distance in the image plane between rays of different wavelength from the same object point.

lateral color see lateral chromatic aberration

lateral force microscopy

A mode of atomic force microscopy in which the the lateral (frictional) forces between a moving probe and a specimen are measured and mapped. In LFM the twisting of the cantilever as the tip moves in contact mode over regions of variable height and composition is measured with a split (quadruple) photodiode detector.

lateral magnification

Linear magnification perpendicular to an optical axis. In a ray diagram, lateral magnification is the ratio of image to object heights above the axis.

lateral spherical aberration

The height above the optical axis where a marginal ray intersects the paraxial focal plane.

lateral spring constant see spring constant

lattice image

Any high-resolution transmission electron microscope image of a thin crystal showing fringes whose spacing is related to the crystal periodicities.

Laue zones

The zones in reciprocal space that describe crystallographic properties in electron and X-ray diffraction

The zero-order Laue zone is that plane of reciprocal lattice points that passes through the origin of the Ewald sphere of a diffracting crystal. The lattice layer immediately above is the first-order Laue zone, the next the second-order Laue zone, and further layers are called higher-order Laue zones.

law of reciprocity

A law relating the exposure of photographic film to the ambient light conditions: film exposure will be the same when the product of light intensity and time is constant.

law of reflection

The law of reflection states that when the incident ray and the reflected ray are in the same plane, the angle of incidence equals the angle of reflection.

law of refraction

The law of refraction states that when the incident ray and the refracted ray are in the same plane, the angle of incidence i is related to the angle of refraction r by the formula:

$$n_1 \sin\theta i = n_2 \sin\theta r$$

where n_1 and n_2 are the refractive indices of the media on the incident and refracting sides of the interface respectively.

lead citrate

A stain widely used to contrast thin plastic sections of biological specimens for transmission electron microscopy.

Lead citrate is often used after staining with uranyl acetate which may act as a mordant for lead. Common formulations are: Venables and Cogeshall lead stain; Reynolds' lead citrate.

left-circularly polarized light

Circularly polarized light in which the electric field vector rotates anticlockwise (as seen by an observer looking towards the source).

lens

A refracting device that focuses or modifies the path of light or radiation. In light microscopes, lenses are typically made of glass. In electron microscopes, lenses are magnetic or electrostatic.

lens coating see antireflection coating
lens coils see windings

lens current

The electric current flowing in the winding of a magnetic lens. The lens current determines the focal length of the lens.

lens equation see lensmaker's equation, Newton's lens equation, thin-lens equation
lens excitation see lens current

lens gap

The space between the polepieces of a magnetic lens, often filled with a non-magnetic spacer. The lens gap is the region where the magnetic field is most concentrated.

lensless microscope

A microscope that has no lenses. Types of lensless microscopes include: projection microscope, scanning probe microscopes, 3D atom probe. Lensless microscopes do not suffer from the aberrations of lens-based microscopes.

lensmaker's equation

For a thin glass lens surrounded by air:
$$1/s_o + 1/s_i = (n - 1)(1/r_1 - 1/r_2)$$
where n is refractive index of glass, s_o and s_i are object and image distances from lens vertices respectively, and r_1 and r_2 are the radii of curvature of the lens surfaces.

lens of eye see eye lens
lens rotation centering see current centering

lens vertex
The point of intersection of a surface of a lens with its optical axis.

lenticular autostereoscopy
An autostereoscopic technique in which an array of vertical cylindrical lenses (lenticules) is placed over a vertically interlaced stereo pair of images; the lenses refract light from only one image into each eye.

lenticular eyepiece see **expanded-pupil eyepiece**
Lichte focus see **Gabor focus**

lift mode, in SPM
A mode in scanning probe microscopy, used after specimen topography has been determined, in which the probe is lifted and scans the specimen at a specific height above its surface.

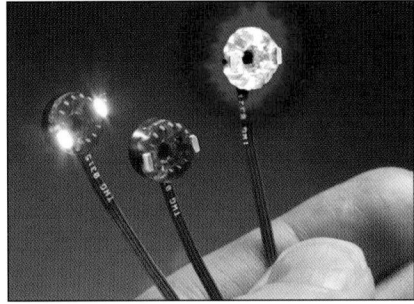

light
The wavelengths of electromagnetic radiation that are detectable by the human eye, typically ranging from 400 nm (violet) to 700 nm (red). Light is also loosely used to describe ultraviolet and infrared radiation, to describe all forms of electromagnetic radiation, or as a synonym for photons.

light chopper
A device that temporarily interrupts the passage of a beam of light, e.g. a rotating mask.

light-emitting diode
A semiconductor that emits light by electroluminescence. A light-emitting diode is a forward biased p-n junction composed of compounds of gallium and indium. When a voltage is applied electrons cross the junction from the n- to p-type material and combine with holes producing electroluminescence. The wavelength of emitted light is a function of the LED material: e.g. gallium nitride (blue); gallium phosphide (red or green), gallium arsenide (infrared).

Light emitting diodes. Top: custom LEDs. Bottom: LEDs on a circuit board. Courtesy of StockerYale Inc.

light guide
A narrow channel that transmits light between optical devices, e.g. a fiber optic.

light microscope

A microscope that uses visible light as the source of illumination. Without qualification, the term light microscope is typically used to refer to an upright compound light microscope. The key components of a light microscope are a light source, a condenser, a stage, one or more objective lenses, a tube with intermediate optical accessories, and a viewing head with one or two eyepieces. These components are attached to a rigid body that carries the focusing mechanisms, accessories and electronics required for use of the microscope. In standard brightfield imaging mode using Köhler illumination, light from each point in the front focal plane of the condenser passes through the condenser lens and travels through the specimen as a parallel beam to the objective lens which brings all beams to a focus in its back focal plane, where a diffraction pattern of the specimen may be observed. The light beams then diverge to form a image in the primary or intermediate image plane, located at the front focal plane of the eyepiece. The eyepiece magnifies the intermediate image producing parallel beams from each point in the intermediate image that converge at the eyepoint; this is where an observer positions their eye lens which focuses the beams onto the retina, creating a virtual image that appears to lie about 25 cm in front of the observer.

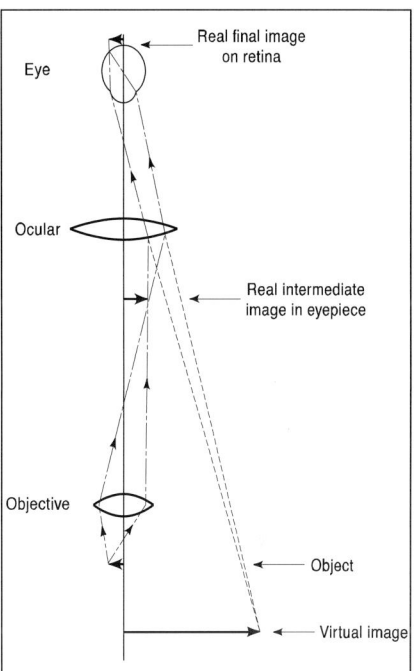

Principle of image formation in a light microscope. From D. Murphy © John Wiley and Sons, Inc.

light microscopy

The use of a microscope with visible light as the source of illumination.

light pipe see fiber optic

light scrambler

A device that transmits a spatially uniform beam of light, such as a fiber-optic cable or ground glass diffuser.

light shutter see shutter

limb see microscope limb

limit of detection see detection limit

limit of resolution see resolution

line (to line) resolution

The smallest distance between two clearly resolvable lines in an image.

linear magnification

The ratio of a linear distance in an image to the corresponding distance in an object.

linear polarizer

A polarizing filter that transmits linearly polarized light.

linearity

The relationship between the input to the output signals of an imaging device. For quantitative microscopy, linearity should be perfect, i.e. the difference in intensity of points in the specimen should be the same in the image.

linearly polarized light

Light waves with a single plane of vibration of the electric field vector.

liner tube

The inner lining of the column of an electron microscope. The liner tube is made of aluminum or brass and is designed to minimize the volume of the column that has to be evacuated, absorb X-rays and produce secondary photons that can be blocked by the lead shielding, and reduce the number of O-rings required between parts of the column.

liquid crystal

An organic substance that is liquid but has crystalline order to its long molecules. The orientation of liquid crystal molecules can be controlled with applied electric fields as in a liquid crystal display. Nematic liquid crystals have parallel orientation but are randomly arranged; cholesteric crystals are arranged in layers with the molecular axes parallel to the layers; smectic crystals are arranged in layers with molecular axes perpendicular to the layers.

liquid-crystal display

A image display device containing an array of pixels each comprising a layer of liquid crystals sandwiched between two electrodes and crossed polarizers, and mounted over a light source. Light is polarized by the first polarizer, optically rotated by the liquid crystals, and is transmitted by the upper polarizer. When an electric field is applied across the liquid crystal layer the crystals can be variably reorientated, changing the degree of optical rotation and hence the amount of light that is blocked by the upper polarizer. In a color LCD each pixel comprises three liquid crystal cells with red, green or blue filters.

liquid-crystal tunable filter

A tunable interference filter consisting of a series of waveplates, each comprising a quartz retardation plate and a liquid crystal layer sandwiched between polarizers. The amount of retardation of each waveplate and hence wavelength-selective blocking is determined by the voltages applied to the liquid crystal layer.

liquid light guide

A light guide that contains a liquid, e.g. alcohol or water, as the transmitting material.

liquid-metal ion source

A source of metal ions used in a focused ion-beam microscope. A liquid-metal ion source consists of a heated tungsten filament tipped with a cone of liquid metal, typically gallium or gold, and a highly negatively charged extraction anode which produces field emission of metal ions.

liquid nitrogen-free detector

Any type of semiconductor detector that operates efficiently without cooling to liquid nitrogen temperatures.

lithium-drifted silicon detector

A silicon semiconductor doped with lithium, commonly used in X-ray spectrometers. A Si(Li) detector is a reverse-biased p-i-n diode with a 3-mm thick Li-drifted intrinsic region. The excitation energy is ~3.8 eV. Incident X-ray photons generate charge carriers, in proportion to photon energy, that are collected by thin metal electrodes on either side of the detector.

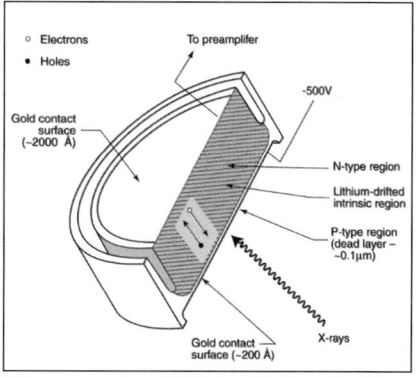

Schematic of lithium-drifted silicon [Si(Li)] detector. Courtesy of Thermo Electron Corporation.

197

livetime

The time that a detector is collecting signals.

longitudinal color see **chromatic aberration**
longitudinal magnification see **axial magnification**

longitudinal spherical aberration

The distance between the focal points of marginal and paraxial rays.

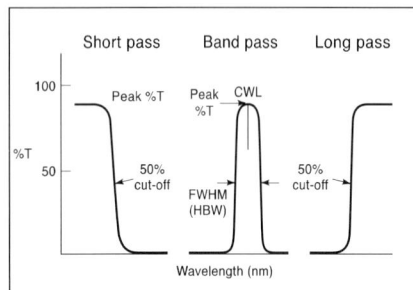

Principle of longpass filter. From D. Murphy © John Wiley and Sons, Inc.

longpass filter

A filter that transmits wavelengths longer than a certain specification and blocks shorter wavelengths.

long working-distance condenser

A condenser with a working distance of several centimeters, typically used in inverted light microscopes to allow access to the specimen.

long working-distance objective

A light microscope objective with a working distance of many millimeter typically used for the examination of thick specimens or to allow space for manipulation of the specimen.

loop see **transfer loop**

Lorentz lens

A magnetic lens with a weak magnetic field, used in Lorentz microscopy of magnetic specimens.

Lorentz microscopy

The use of a transmission electron microscope to image the domain structure of magnetic materials. In Lorentz microscopy the specimen should be mounted in a region of the column that has minimal magnetic fields: in a specialised stage between the condenser and objective lenses, in a Lorentz lens, or by turning off the objective lens and operating the microscope in low magnification mode. Contrast arises from interference of electron waves refracted towards or away from each other in adjacent domains, producing a Fresnel fringe. The two domains act rather like a biprism.

loss of mass

A consequence of radiation damage to resin sections and thin foils examined in an transmission electron microscope. Loss of mass can be caused by volatilization of organic components from resin and polymer sections, and sputtering of surface atoms from foils.

low-angle diffraction see small-angle diffraction

low-dose imaging

The use of low-intensity electron beams to record structural or compositional information in a transmission electron microscope. Low-dose imaging is a requirement for cryoelectron microscopy of native, hydrated biological samples where the maximum dose should not exceed 100 electrons nm^2 s^{-1}, and the total specimen dose should be less than 2,000 electrons.

low-dose kit

A software program for cryo-electron microscopy which controls low-dose imaging of beam-sensitive specimens. The software automatically controls specimen position, focus, shutter and exposure settings, allowing a specimen to be searched at low magnification, focused at high magnification on an area of the specimen adjacent to that of interest, and photographed by translating the beam to the area of interest for a low-dose exposure of film or CCD camera.

low-energy electron diffraction

A form of backscattered electron diffraction using low energy electrons, typically 20-200 eV.

low-energy electron microscope

A electron microscope that operates at low accelerating voltages (~10 V) used primarily for the in-situ, real-time study of dynamic processes in materials by reflection of low-energy electrons. The low-energy electron microscope has a conventional high-voltage electron gun but the beam is decelerated before reaching the stage. Primary electrons reflected by the surface of the specimen are collected to form an image or diffraction pattern.

low-energy electron point-source microscope

A point projection electron microscope that uses a low-energy (20-200 eV) electron beam.

low-loss region

A region in an electron energy-loss spectrum formed by electrons with energy losses up to about 50 eV. The low-loss region contains plasmon peaks and inter-and intraband transition peaks.

low-pass filter

An image filter that reduces high spatial frequency details, smoothing the image.

low-temperature embedding chamber see cryochamber

low vacuum

A gas pressure in the range 10^{-1} to 10^{-4} pascals.

low-vacuum scanning electron microscope see environmental SEM, variable-pressure SEM

low-voltage electron microscope

1. A conventional light microscope in which the stage and illumination system are replaced with a miniature transmission electron microscope column. The low-voltage electron microscope operates at accelerating voltages of 1-5 kV. The electron optics project an image of the specimen onto a YAG screen and the light microscope is used to magnify the primary image.
2. A scanning or transmission electron microscope that typically operates at accelerating voltages in the range 1-5 kV.

low-voltage scanning electron microscopy

The use of a scanning electron microscope operated with low accelerating voltages (typically 1-5 kV) for the study of uncoated or beam-sensitive organic specimens such as intracellular structures and polymer films. Low voltages reduce charging of uncoated specimens.

lumen

SI derived unit of luminous flux. Symbol lm. One

lumen equals the light emitted into a unit solid angle (steradian) by a point source of one candela.

luminance

A photometric quantity used to characterize the luminous flux arriving, passing through or leaving a unit area. Symbol L ; units: lumen per steradian per square meter.

luminescence

The emission of light by a material as a result of excitation by any form of energy. Short-term luminescence is called fluorescence; long-term luminescence is called phosphorescence. The different types of luminescence are specified by adding a prefix indicating the source of excitation energy: for example, photoluminescence is excitation by light, cathodoluminescence is excitation by electrons.

luminosity

The apparent brightness of an object or color.

luminous energy

A photometric quantity used to characterize the energy of electromagnetic radiation. Symbol Q; units: lumen second (lm s).

luminous flux

A photometric quantity used to characterize the rate of flow of electromagnetic radiation. Symbol F; units: lumen.

luminous intensity

A photometric quantity used to characterize the intensity of electromagnetic radiation. Symbol I; units: lumen per steradian (lm sr^{-1}).

lux

SI derived unit of illuminance. Symbol lx. One lux equals one lumen per square meter.

M

Mach-Zehnder interferometer

An interferometer consisting of two beamsplitters and two mirrors placed at the corners of a square. Light enters the inteferometer and is split into two beams by the first beamsplitter; these beams are reflected by the mirrors to the second beamsplitter that recombines the beam, producing an interference pattern if an object is placed in one of the pathways.

machine vision

The use of high-speed video cameras and image analysis software to locate, measure and control a particular activity, e.g. for quality control of objects on a production line.

macrophotography see photomacrography

macroscope

An optical system used to examine large objects.

macula

A 3-mm diameter depressed region of the retina containing a higher number of cones than rods.

magnesium fluoride coating

A thin film of magnesium fluoride applied to the surfaces of lenses to reduce the reflection of light.

magnetic field cancelation

The reduction of the effects of stray magnetic fields on the performance of electron microscopes. A magnetic field cancelation system has sensors close to the stage measuring x, y and z orientated magnetic fields, and Helmholz cables in similar orientation placed on the walls of the microscope room. A controller reads the magnetic fields at the stage and uses the information to create magnetic fields of opposite phase in the Helmholz cables to reduce aberrations.

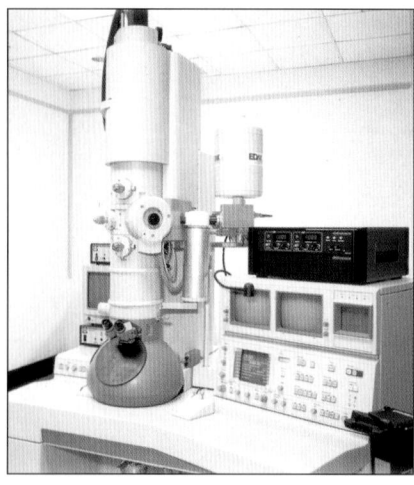

A magnetic field cancelation system in a room containing a transmission electron microscope. Courtesy of ETS Lindgren.

magnetic field vector

A vector that describes the direction and amplitude of vibration of the magnetic field along the axis of an electromagnetic wave.

magnetic force microscopy

The use of a scanning probe microscope to map and measure the magnetic fields emanating from a specimen. In magnetic force microscopy a tip coated with a ferromagnetic film is used to detect magnetic fields by changes in cantilever deflection or resonant frequency. Typically a two-pass method is used: first the topography of the specimen is imaged; the tip is then moved away from the specimen and a second scan measures magnetic fields.

Magnetic force microscopy of domain structure of film of yttrium iron garnet. H = 31 Gauss. Scan: 60x60 μm. Courtesy of NT-MDT and A. Temiryazev and M. Tikhomirova, Institute of Radioengineering & Electronics RAS, Fryazino, Russia.

magnetic lens

A lens that focuses charged particles using electrically induced magnetic fields, the principal probe-forming and magnifying device used in electron microscopes. A round magnetic lens comprises a coil of copper wire, the winding, and a soft-iron yoke or shroud that encases the winding and is terminated by polepieces that surround a central hole, the lens bore. The lens contains channels for cooling water. As current passes through the lens it magnetizes the shroud, producing a strong magnetic field between the polepieces. If the bore-radius is the same on both sides of the lens gap separating the polepieces, the lens is said to be symmetric, since the magnetic field on the axis will be symmetric about its mid-point; otherwise, the lens is asymmetric. Changing the lens excitation affects the focal length and hence the behavior of the lens.

magnetic prism

A curved magnetic lens used in electron spectrometers. A magnetic prism produces a uniform magnetic field between two parallel faces of the prism; electrons passing through the prism experience a force perpendicular to the field and their original direction of travel and are bent through a large angle, typically 90°. The prism disperses electrons according to energy.

magnetic resonance force microscopy
The use of a scanning probe microscope with a magnetic tip mounted on an ultrasensitive cantilever to image molecular atomic structure by detecting electron spin behavior in a nuclear magnetically resonating sample.

magnetic-sector prism see **magnetic prism**

magnetic-sector spectrometer
An electron spectrometer with one or more magnetic prisms.

magnetic shielding see **Faraday cage**
magnetic shroud see **magnetic yoke**

magnetic yoke
The metal casing of a magnetic lens that becomes magnetized when current flows through the winding.

magneto-optical Kerr effect
The change in state of polarization of light upon reflection from a magnetic specimen.

magneto-optical microscopy
The use of a polarized light microscope to study the magneto-optical Kerr effect.

magnification
The degree by which dimensions in an image are enlarged compared with the same dimensions in the object.

magnifier see **magnifying glass**

magnifying glass
A simple or compound converging lens placed between an object and the eye that forms a magnified virtual image.

magnifying power see **angular magnification**

mandoline energy filter
An in-column electron spectrometer having three magnetic prisms arranged in the form of a

mandoline: an on-axis prism deflects the beam towards two sector prisms that return the beam to the optic axis through the first prism.

mapping

The generation of an image that shows the spatial relationships of a property of a specimen.

mapping element see dot mapping

marginal ray

A ray that passes from the object point to the margin of the entrance pupil.

marking objective

A rotatable marking device with a diamond tip that can be placed in the objective turret of a light microscope and used to enscribe glass coverslips or hard specimens.

mass-absorption coefficient

A coefficient used in X-ray microanalysis that describes the reduction in the initial intensity of emitted X-rays as they pass through a specimen or detector. The mass absorption coefficient μ/ρ has units of area per mass (usually the non-SI units $cm^2\ g^{-1}$).

mass mapping

The production of an image in which intensity is related to specimen mass, typically using scanning transmission electron microscopy.

mass thickness

The combination of density and thickness which is a measure of the ability of a specimen to absorb X-rays and scatter electrons. Mass thickness affects the intensity in X-ray spectra and the contrast in electron microscope images.

mass-thickness contrast

Contrast that derives from the use or exclusion of scattered beams in image formation. In mass-thickness contrast imaging in a transmission electron microscope the objective aperture is used to block electrons scattered by regions of the specimen having high density and atomic number Z of constituent atoms. Mass thickness contrast is the primary mechanism of image formation in stained biological and amorphous materials specimens.

mathematical morphology

A branch of image processing based on two nonlinear operations, erosion and dilation, widely used for image enhancement and image analysis.

matrix optics

A method in optics for tracing rays through an optical system using matrix methods.

mean free path

The average distance traveled by radiation between successive scattering events. Symbol λ> The mean free path is the reciprocal of the interaction cross-section.

mechanical advance microtome

A microtome with a system of mechanical gears to advance the specimen towards the knife.

mechanical pump see rotary pump
mechanical tube length see tube length

median filter

An image filter that replaces a specified pixel value with the median value of the pixel and its neighbors, smoothing the image.

meniscus force

A force caused by the surface tension of water molecules. Meniscus force is a potential problem when operating a scanning probe microscope in air and is caused by condensation of water on the specimen surface leading to wetting of the probe.

meniscus lens

A lens with one convex and one concave surface, shaped like a meniscus.

mercury arc lamp

An arc lamp that contains liquid mercury and small amounts of either argon or xenon. The gases create an initial plasma that vaporize the mercury. A mercury arc lamp produces a continuous spectrum of radiation from UV to IR with intense

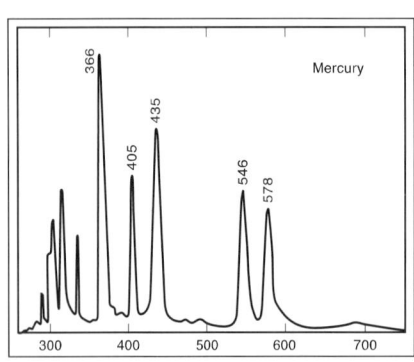

Emission spectrum of mercury arc lamp. From D. Murphy © John Wiley and Sons, Inc.

peaks in the UV and visible regions at around 337, 365, 405, 435, 546 and 578 nm, which can be used for the excitation of fluorophores.

mesa

A flat-topped steep-sided projection, the ideal shape of a block for sectioning on a microtome.

mesh grid

An electron microscope support grid with a hexagonal or rectangular mesh, the most common type of grid used for the examination of thin sections.

metal cathode see target
metal coating see replication

metal-halide arc lamp

A mercury arc lamp containing halides of sodium and rare earth metals such as Sc, Tl, In, and Dy, that generate an emission spectrum similar to sunlight.

metal-mirror cryofixation

A method of cryofixation in which the specimen is frozen by rapid contact with a cooled, polished metal surface.

metal-mirror cryofixation device

A device used for metal mirror cryofixation. In a typical device the specimen is placed on a pneumatically or gravity driven specimen holder and is then slammed onto a liquid nitrogen or liquid helium-cooled copper disk. A damping mechanism prevents specimen compression and bounce. Metal mirror cryofixation can produce vitrification of specimens 10-20 μm thick.

metal-oxide semiconductor

A semiconductor capacitor comprising a metal electrode, a layer of silicon dioxide and a layer of silicon. The MOS is the basic building block of CMOS and CCD devices.

metallurgical microscope

A light microscope designed for the study of metals. A metallurgical microscope is typically equipped with an epi-illuminator for the

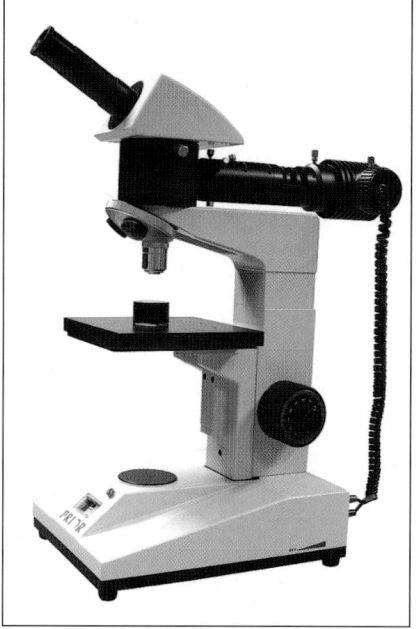

Metallurgical microscope. Courtesy of Prior Scientific Instruments.

examination of opaque specimens.

meter

SI base unit of length. Symbol m. One meter is the distance travelled by light in a vacuum during one 299,792,458th of a second.

methacrylate

A synthetic acrylic resin formed by polymerization of methacrylic acid or its esters.

metrology

The science of measurement.

mica

A silicaceous material, such as muscovite, with closely spaced crystalline sheets. Mica is easily cleaved to produce a planar, featureless substrate for the manufacture of carbon support films and for replication of particles and macromolecules.

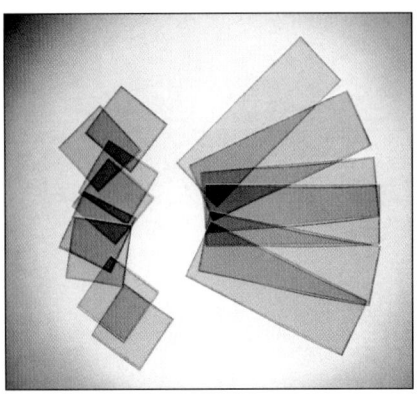

Mica sheets. Courtesy of Agar Scientific.

Michelson interferometer

A dual beam interferometer consisting of a beam splitter and two mirrors. Light entering the interferometer is split into two orthogonal beams by the beam splitter; one beam passes to a fixed mirror and the other to the reference mirror which can be moved axially. After reflection by their respective mirrors the beams are recombined by the beamsplitter generating an interference pattern due to optical pathlength differences between the two beams.

microanalysis

The chemical analysis of very small regions of a specimen using a microscope. Microanalysis is usually performed with focused beams of electromagnetic radiation such as ultraviolet, visible light, infrared and X-rays, or with charged particles such as ions and electrons. In the electron microscopy community, microanalysis refers to the use of an electron microprobe to produce X-rays that are analyzed by spectrometry. The key feature of microanalysis is its spatial resolution, unrivalled by any other analytical technique, which is a function of the electron probe size at the specimen; for thin sections this can be in the nanometer range with field emission guns.

microcalorimeter detector

A heat-sensitive device used for X-ray detection. A microcalorimeter has an X-ray absorber, a thermister and a heat sink, typically maintained at ~0.06K. When an X-ray is absorbed it generates photoelectrons that produce a temperature rise on the order of a few millikelvin that is proportional to the energy of the incident X-ray.

microchannel plate

An electron multiplying device with parallel continuous dynodes that preserves the spatial resolution of the input image. A microchannel plate is a thin sheet of closely packed, vertical micrometer-diameter lead-glass capillaries, chemically treated to enhance secondary electron emission. Photons or charged particles entering the channels strike the walls generating secondary electrons that are accelerated by an applied voltage through the capillaries, generating further secondary electrons by avalanche multiplication.

microdensitometer

A device that measures optical density or contrast in a small area of an object or a photographic image.

microdiffraction see nanodiffraction
microdiffraction pattern see nanodiffraction pattern

micrograph

A recorded image produced on a microscope.

micrography see photomicrography
microlever see cantilever

microinjector

A device for the pressurized delivery of small volumes of material through a microcapillary.

microlens

A very small converging lens. Microlenses are used to focus light onto photodiodes of charge-coupled devices and onto pinholes of spinning disks.

micromanipulator

A device for the manipulation of specimens with microneedles.

micrometer

A device for measuring small distances.

micrometer-screw eyepiece

An eyepiece with an integral micrometer screw which moves one reticle with respect to another for measurement of image distances.

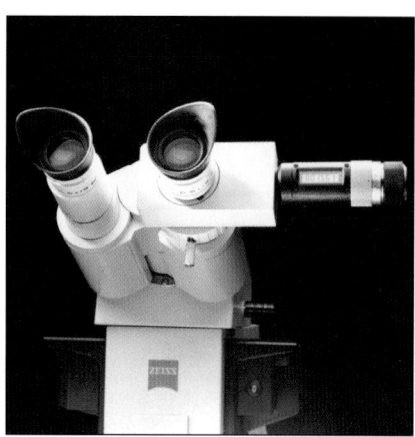

Micrometer-screw eyepiece. Courtesy of Carl Zeiss Ltd.

micron

An obsolete term for micrometer.

microphotography

The photography of large objects to produce a demagnified image, such as the recording of documents and newspaper pages on microfiche.

microprobe mode

The use of a micrometer-diameter parallel beam of electrons in an electron microscope.

microprobe see electron microprobe

microprojector

A light microscope with optics to project the image onto a built-in or remote screen.

micro-Raman microscopy

The use of a Raman microscope to identify the spatial distribution of the chemical components of a specimen.

microscope

An instrument that forms a magnified image of a specimen. The type of microscope is usually denoted by the use of one or more prefixes or qualifiers, e.g. electron microscope, polarized light microscope, scanning probe microscope.

microscope base

The base of the stand of a light microscope, typically housing the illuminated field diaphragm, a mirror to direct light to the condenser, neutral density filters, and many of the mechanical and electrical controls for operation of the microscope.

microscope shutter see **shutter**
microscope stage see **stage**

microscope stand

The chassis of a light microscope, comprising base, substage, stage and limb, that carries the mechanical, electronic and optical components.

microscope tube

The hollow tube through which light travels between the objective lens and the eyepiece(s) of a light microscope. The term originates from the straight tube used in early microscopes that held the objective nosepiece at one end and the eyepiece at the other. Modern microscopes with infinity-corrected optics are modular and the tube is now typically formed by a combination of several separate, optically coupled components placed between the objective and eyepiece such as the body tube, the intermediate tube, eye-level risers, the viewing head and the eyepiece tube(s).

microscopy

The use of a microscope to produce an image or to investigate the properties of a specimen.

microtome

An apparatus used to cut sections of a specimen. All microtomes have mechanisms for moving the block past a knife during the cutting stroke, for retracting the knife or displacing the specimen during the return stroke, and for advancing the specimen towards the knife (or *vice versa*) by a preset distance for each cutting cycle.

microtome chuck

The device that holds the block on a microtome. For ultramicrotomy specialized chucks are available for differently shaped blocks, e.g. flat or cylindrical, and different types of specimen, e.g. unembedded or frozen.

microtome knife

The blade used to cut sections on a microtome. Microtome knives may be made of steel, glass, diamond or sapphire.

microtomography

Tomography of small objects.

microtomy

The cutting of sections of a specimen on a microtome.

microwave

Electromagnetic radiation with wavelengths in the range 1 mm to 0.3 m.

microwave fixation

The use of microwaves to assist the chemical fixation of a specimen. Microwave irradiation induces vibrations and localized heating in the specimen leading to more rapid penetration of fixative and assisting cross-linking of components, allowing complete fixation in minutes. Specimens should be small (<3 mm) and specimen temperature must be monitored to avoid heat damage.

microwave microscopy

The use of a microscope with microwave radiation as the source of illumination.

microwave processing

The use of microwaves to enhance the effectiveness of a specimen processing procedure, such as fixation.

mid infrared

Electromagnetic radiation with wavelengths in the range 3-6 μm.

Mie scattering

The scattering of a light wave by objects larger than its wavelength. Mie scattering is generally small angle, forward scattering.

millibar see bar

milling

The removal of surface material by physical means. Milling techniques include grinding, polishing, ion-beam etching.

minicondenser lens

A condenser lens placed inside objective lens of an electron microscope.

minilens see miniscondenser lens

minimum contrast focusing

A procedure used to assist focusing of a specimen in a transmission electron microscope: the objective aperture is removed and the plane of focus that gives minimum contrast in the image is selected.

minimum detectable mass

The minimum number of atoms detectable by X-ray microanalysis in the analysis volume of a specimen.

minimum-dose imaging see low-dose imaging

minimum fold

A fold or crease that projects normal to the surface of a section. Minimal folds may be used to measure section thickness if the fold diameter is twice the section thickness.

minimum mass fraction

The minimum concentration of an element (in wt% or parts per million) that can be detected by X-ray microanalysis in the analysis volume of a specimen.

minimum resolvable distance

The minimum distance between points in a specimen that can be clearly distinguished in an image.

mirror block see filter cube

mirror objective

A light microscope objective that uses mirrors to form an image.

mirror stereoscope

A stereoscope having two sets of parallel 45°-inclined facing mirrors, used for viewing large stereo pairs. Light from each image is reflected by the larger outer mirror onto the inner smaller mirror which reflects the image into one of the viewer's eyes, with or without the aid of an additional magnifying lens.

mobile microscope see field microscope
modified Wollaston prism see Nomarski prism

modulation

The process or result of changing the parameters of an image or signal.

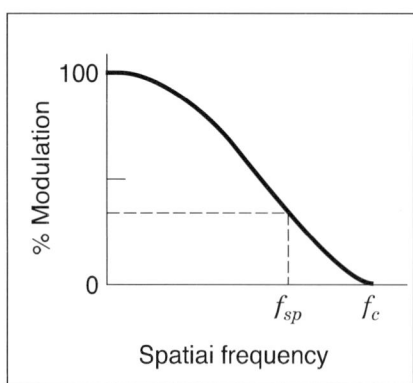

Plot of percentage modulation or contrast vs spatial frequency. fsp: 30% modulation. fc: cut-off frequency. From D. Murphy © John Wiley and Sons, Inc.

modulation contrast microscopy

A class of light microscope imaging techniques that use oblique illumination of the specimen and spatial filters in the back focal plane of the objective, modulating the zero-order of the diffraction pattern. Types of modulation contrast techniques include: Hoffmann modulation contrast, Ellis single side-band edge enhancement, Zeiss variable relief contrast and Olympus relief contrast.

modulation transfer function

A measure of the ability of an optical system or imaging device to transfer signals from object to image as a function of the spatial frequency of the signal. The modulation transfer function is the ratio of image modulation to object modulation at all spatial frequencies.

moiré fringes

The dark lines in a moiré pattern

moiré pattern

A pattern seen when two slightly inclined grids or sets of parallel lines are superimposed.

molar extinction coefficient, of fluors

A measure of the ability of a fluorophore to absorb photons. Symbol ϵ. The coefficient is usually given as the absorption maximum at a specified wavelength.

monitor

An electronic device that displays an image. Most

monitors used in imaging systems use a cathode
ray tube or liquid crystal display.

monochromat
1. A light microscope objective with minimal
chromatic aberration at one wavelength.
2. A lens with similar specification.

monochromatic aberrations
Aberrations of lenses that are independent of
wavelength. Monochromatic or Seidel aberrations
are caused by departures of lens properties from
the paraxial approximation. The monochromatic
aberrations include: astigmatism, coma, distortion,
field curvature and spherical aberration.

monochromatic objective see monochromat

monochromatic radiation
Radiation with a single energy, frequency or
wavelength.

monochromator
A device that produces radiation with a single
wavelength or energy from a polychromatic
source.

Monte Carlo simulation
The modeling of the trajectories or mean free paths
of incident electrons in a specimen. A Monte Carlo
simulation uses a random number generator to
simulate elastic and inelastic scattering of
electrons in the specimen, and is used as an aid to
understanding beam spreading and the interaction
volume.

mordant
A chemical that enhances a staining process by
acting as a catalyst or intermediary.

motorized stage
A stage with stepper motors to allow remote
control of movement in x, y and z directions.

mountant see mounting medium

mounting medium

A medium in which specimens are embedded for examination by a microscope, usually placed between a slide and coverslip.

mounting thread

The screw thread used to attach light microscope objective to the turret. The standard mounting thread is ~20.3 mm diameter, which allows exchange of objectives between microscopes. Some manufacturers now use larger mounting threads of 25 or 32 mm diameter.

multichannel analyzer

An electronic component used in an X-ray spectrometer that assigns charge pulses into a computer memory bank with up to 2048 locations (channels) corresponding to photon energy and displays the data as an X-ray spectrum.

multichannel unmixing see spectral unmixing

multielement lens

A compound lens formed by cementing several lenses together. Multielement lenses are commonly used in light microscope condensers, objectives and eyepieces for the correction of aberrations.

multihead microscope

A light microscope with a primary viewing head containing beamsplitters that direct light to one or more secondary viewing heads, typically used for teaching purposes.

multilayer antireflection coating

An antireflection coating comprising up to one hundred layers of alternating high and low refractive index materials.

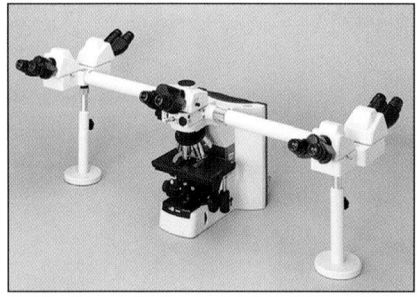

Multihead teaching microscope. Courtesy of Nikon.

multi-immersion objective

A light microscope objective designed for use with different types of immersion media, such as oil and water.

multiphoton microscopy

The use of two- or three-photon excitation in fluorescence microscopy.

multiple scattering

The scattering of an an electron or photon as a result of consecutive interactions with more than 20 atoms.

multiple specimen holder

A specimen holder that has receptacles for two or more electron microscope grids.

multiple wavelength anomalous dispersion see anomalous dispersion

multipole lens

An electron lens with several magnetic poles arranged in a circle around the optic axis, typically used for aberration correction in electron microscopes and spectrometers. Multipole correctors are typically configured as quadrupole, sextupole, or octopole.

multislice method

A mathematical procedure that models the transmission of an electron beam through a specimen by dividing the specimen into a set of slices normal to the beam.

multispectral imaging system

An imaging system that outputs a spectral image.

myopia

An aberration of the eye in which parallel rays from distant objects are focused in front of the retina; near objects are focused correctly. Myopia can be corrected with a negative spectacle lens.

N

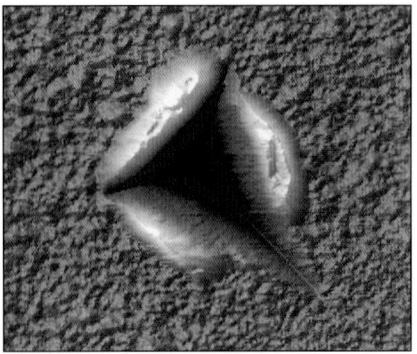

AFM image of nanoindentation of glass with a diamond probe. *Microscopy and Analysis* 2004.

nanodiffraction

A general term for a family of electron diffraction techniques using nanometer-sized electron probes.

nanodiffraction pattern

A diffraction pattern produced using a nanometer-sized electron beam.

nanogold

Proprietary name for 1.4-nm diameter gold particles. Nanogold particles comprise gold atomic clusters surrounded by an organic shell that can be covalently bound to other probes. The small size of nanogold particles facilitates diffusion into cells and tissues but makes them difficult to resolve, so they are often enhanced by autometallography.

nanoindentation

The measurement of specimen hardness by pushing a sharp probe into the surface of the specimen. Hardness H is defined as the maximum contact pressure: $H = F/A$ where F is maximal load and A is area of impression.

nanolithography

The fabrication of nanometer-sized structures on a specimen by the addition, modification or removal of material.

nanometer see meter

nanoprobe mode

The use of a nanometer-diameter parallel beam of electrons in a (scanning) transmission electron microscope, for nanodiffraction and imaging.

nanotube

A cylinder composed of fullerenes (polyhedral clusters of 60 carbon atoms) with a diameter of several nanometers. Nanotubes are used as tips in scanning probe microscopy.

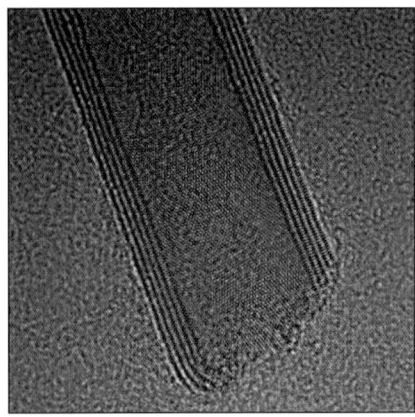

High-resolution TEM image of multiwalled carbon nanotube. Courtesy of FEI Company and Weizmann Institute, Israel.

nanotube tips see **tips, SPM**

nanovid microscopy
The use of video-enhanced high-resolution light microscopy to examine the motion of nanoparticles, e.g. colloidal gold attached to cell surfaces.

narrowband antireflection coating
An antireflection coating that reduces reflections of a narrow range of wavelengths of light.

National Television Systems Committee see **NTSC video standard**

natural light
Light that is polychromatic and randomly polarized.

nearfield diffraction see **Fresnel diffraction**
nearfield optical microscopy see **nearfield scanning optical microscopy**

nearfield optics
The use of optical systems where the object to lens distance is less than one quarter wavelength of light and so images are not subject to the effects of diffraction.

nearfield scanning optical microscope
A type of scanning probe microscope that scans an optical probe in the near field of a specimen to obtain images and data at nanometer resolution. The key components of a near-field scanning optical microscope are: 1. a light source, usually a laser, to illuminate the specimen; 2. a sharp probe that projects or reflects light onto the specimen or collects light emitted by the specimen; 3. a positioning system to control the working distance of the probe; 4. a scanning stage, usually a piezoelectric scanner, to move the specimen relative to the probe at nanometer resolution; 5. an optical system with a detector or spectrometer to receive and analyze the signals emitted from the specimen; 6. a computer to drive the system and process the images; and 7. a robust frame to reduce vibration. There are many configurations of

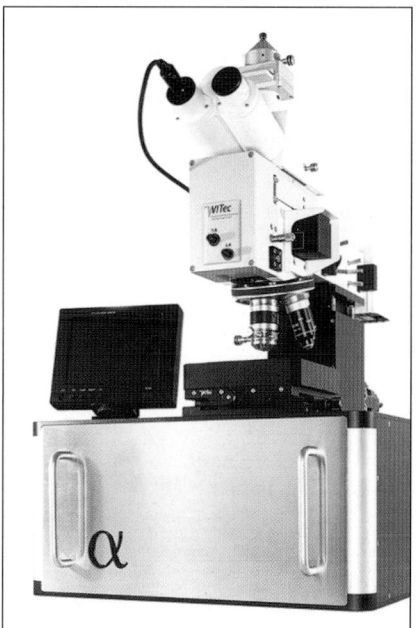

Nearfield scanning optical microscope. Courtesy of WITec.

near-field scanning optical microscopes, and some combine the capabilities of an atomic force microscope. A typical instrument for transmitted light NSOM uses a probe, with a small (~100 nm) aperture at its tip, fiber-optically coupled to a laser. The probe is positioned near the specimen using piezoelectric technology used in other scanning probe microscopes. The stage scans the specimen in raster fashion across the probe. A light microscope objective collects light transmitted by the specimen and passes it to a detector or spectrometer for analysis and image formation. The resolution of an NSOM is in the nanometer range and is a function of the diameter of the probe aperture and the quality of the probe coating.

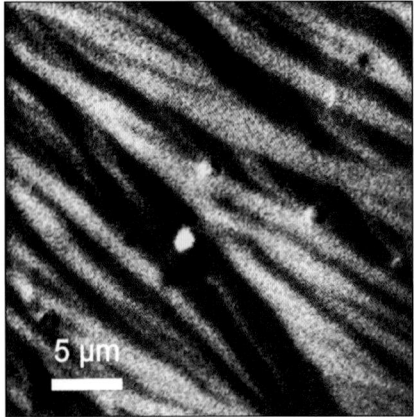

NSOM image of DAN-strands. Courtesy of WITec.

nearfield scanning optical microscopy
The use of a near-field scanning optical microscope.

near infrared
Electromagnetic radiation with wavelengths in the range 700 nm to 1 mm.

near point
The closest point on which the fully accommodated eye can focus. The standard near point is 250 mm.

near ultraviolet
Electromagnetic radiation with wavelengths in the range 300-380 nm, closest to the visible spectrum. Some animals, e.g. bees, and even humans can 'see' near ultraviolet.

nearsightedness see myopia
negative eyepiece see internal-diaphragm eyepiece
negative lens see diverging lens

negative phase contrast
A mode of phase contrast microscopy where objects with a high refractive index appear bright on a darker background in the image. Negative phase contrast light microscopy uses a negative phase plate in the objective lens.

negative phase plate

A phase plate with an annulus formed by a λ/4 phase-retarding dielectric layer coated with an absorbing metallic film. The annulus attenuates the surround waves and adds a λ/4 retardation.

negative replica

That surface of a replica with a topography that is in reverse orientation to that of the specimen. Negative replicas may be used as molds to form a positive replica.

negative stain

A stain that surrounds the surface of a specimen giving additional contrast to the background. Examples of commonly used negative stains in transmission electron microscopy are: uranyl acetate, phosphotungstic acid and ammonium molybdate.

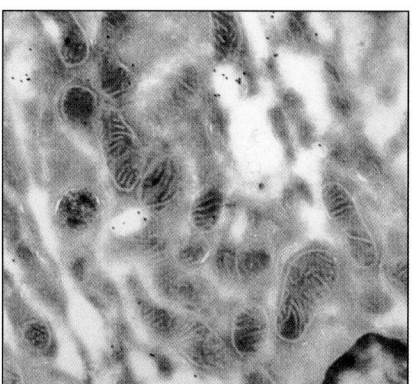

Negatively stained cryosection of entero-cyte showing mitochondrial membranes. Courtesy of Julian Heath.

negative staining

A technique for revealing the structure of specimens in the transmission electron microscope by use of a heavy-metal negative stain. The specimen is immersed in a solution of heavy metal salt which on drying forms an electron-dense cast revealing three dimensional details in negative contrast. Typical applications include ultrastructural studies of cellular organelles, macromolecules, DNA, bacteria, viruses, and small particles.

Nelsonian illumination see source-focused illumination

neutral beamsplitter

A beamsplitter that reflects and/or transmits light of all wavelengths. A neutral beamsplitter is used in reflected light microscopy.

neutral-density filter

A filter that reduces the intensity of all wavelengths of light by absorption or reflection.

newton

SI derived unit of force. Symbol N. One newton is the force required to accelerate a mass of one kilogram at one meter per second per second.

Newton's lens equation

The product of the distances of the object to the front focal plane x_o and of the image from the back focal plane x_i equals the square of the focal length f of a lens: $x_o.x_i = f^2$.

Newton's rings

The concentric interference fringes observed in monochromatic light when a convex transparent object is placed on a flat reflecting surface. The fringes are produced by destructive interference of light reflected from the apposed surfaces and can be used to measure the shape of lenses. The radius of the nth dark ring R_n is given by:

$$R_n = (nr\lambda)^{1/2}$$

where r is radius of curvature of the convex surface and λ is wavelength.

Nicol prism

A polarizing prism comprising two wedges of birefringent calcite cemented together at the hypotenuse so that the ordinary ray is totally internally reflected at the interface and absorbed by an external black coating and only the extraordinary ray is transmitted.

Nipkow disk

A disk with multiple pinholes arranged in a spiral so that during rotation of the disk each hole sweeps across a different region of an object or image, producing a raster scan. A Nipkow disk is used in certain spinning-disk confocal microscopes.

nodal planes

The two planes perpendicular to the optical axis of a lens that are intersected by a ray entering and leaving a lens without deviation in its angle, i.e. with unit magnification. When the lens is surrounded by the same medium, the nodal and principal planes are coincident.

nodal points

The points on the optical axis intersected by the nodal planes.

noise

Any unwanted signal that degrades an image or spectrum. The major sources of noise are the

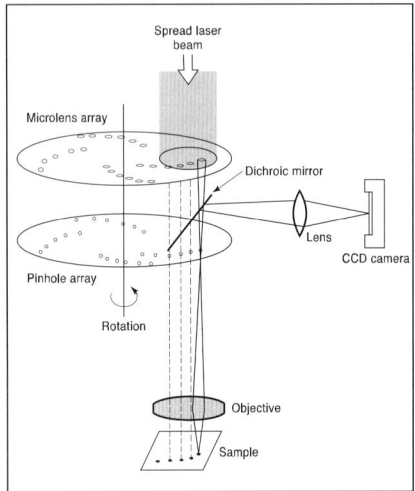

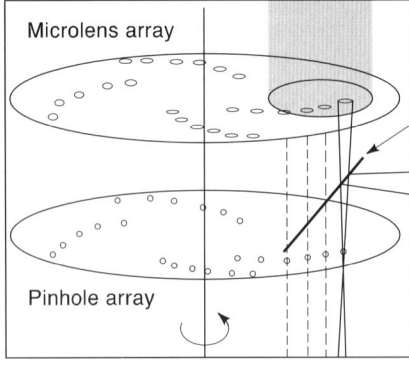

Principle of Nipkow disk in tandem scanning confocal microscope. From D. Murphy © John Wiley and Sons, Inc.

specimen background, shot noise and detector noise (dark current and read-out noise).

Nomarski microscopy see differential interference contrast light microscopy

Nomarski prism

A modified Wollaston prism used as the condenser and objective prisms in most modern differential interference contrast microscopes. In a Nomarski prism the optic axes of the two wedges are at an oblique angle producing an interference plane outside the prism, a property that allows placement of the prism outside the condenser and objective focal planes.

non-contact mode

A mode in scanning probe microscopy in which the probe to specimen distance lies within the attractive regime of long-range van der Waals forces, typically 1-10 nanometers.

non-contact profilometry

Any form of profilometry where there is no direct contact with the specimen.

non-conventional magnetic lenses

Magnetic lenses with an unusual configuration used in specialized electron optical instruments. Examples: pancake lens, snorkel lens, einzel lens.

non-descanned mode

A optical configuration in confocal microscopy in which the light emitted by the specimen does not return through the scanning head but is collected using dichroic mirrors in the emission path.

non-linear microscopy

Any form of microscopy that is used to study non-linear optical phenomena.

non-linear optics

A branch of optics that is concerned with optical phenomena that behave in a non-linear fashion, such as two-photon excitation, second harmonic generation, and anti-Stokes emission.

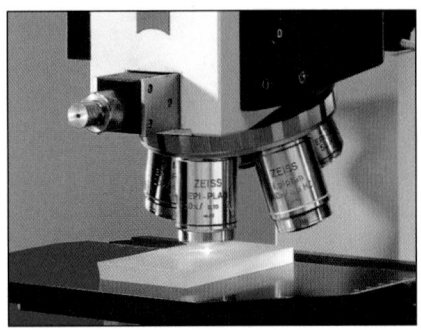

Objective turret or nosepiece of light microscope. Courtesy of Carl Zeiss Ltd.

non-polarized light see **randomly polarized light**

nosepiece

That part of a light microscope which carries the objective(s).

nosepiece turret see **objective turret**

notch filter

A filter that blocks a narrow range of wavelengths of light.

NTSC video standard

The National Television Systems Committee (USA), 525-line, 30 frames per second, color broadcast video standard adopted by the US, Japan and other countries.

numerical aperture

A measure of the light-gathering ability and resolving power of a lens. Numerical aperture NA is given by the expression:

$$NA = n\sin\theta$$

where n is the refractive index of the medium between the specimen and the lens and θ is the aperture semi-angle. The maximum NA of a dry objective is ~0.95. An oil-immersion 100x objective has an NA of about 1.4.

Nyquist criterion

A criterion for the faithful reproduction of object detail by an electronic device such as a CCD camera. The Nyquist criterion states that image resolution must be equal to, or greater than, twice the highest spatial frequency of the detail in the specimen, i.e. an object comprising alternating black and white lines requires at least two pixels of the detector per line.

A
N. A.
= 0,25

B
N. A.
= 0,75

Effect of numerical aperture on light-gathering ability of objective lens. Courtesy of Carl Zeiss Ltd.

O

object

That from which an image is formed.

object distance

The distance on the optical axis between the object plane and the object-side principal plane.

objective see objective lens

objective aperture

1. The aperture of an objective lens.
2. An aperture (diaphragm) beneath the objective lens of a transmission electron microscope.

objective aperture control, of TEM

The mechanism that changes the size of the objective aperture on a transmission electron microscope. The control is located on the side of the column beneath the stage and allows click-stop exchange of apertures and displacement of the aperture from the optic axis for optical alignment and specialized imaging purposes.

objective diaphragm

1. The diaphragm in the back focal plane of a light microscope objective.
2. Synonym for objective aperture of a transmission electron microscope.

objective lens

The primary image-forming lens of a light or electron microscope.

objective lens, of light microscope

The compound lens that forms the intermediate image in a light microscope. The objective lens is a multielement lens housed in a steel barrel with additional optical elements such as phase plates and diaphragms and a screw thread to connect it to the nosepiece.

objective lens, of SEM

The probe-forming lens of a scanning electron

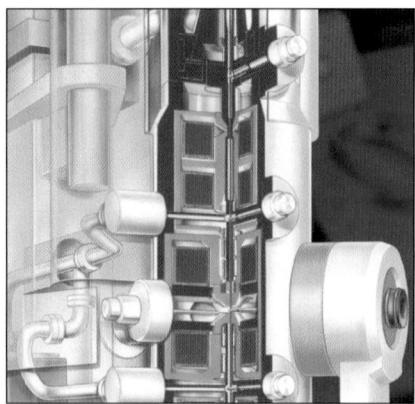

C1 and C2 condenser lenses (top) and objective lens (bottom) of transmission electron microscope. Courtesy of FEI Company.

microscope. Although an SEM does not, in principal, have any post-specimen lenses, the final condenser lens may act as an objective lens since secondary and backscattered electrons may be collected through the condenser lens for imaging purposes.

objective lens, of TEM

The principal image-forming magnetic lens of a transmission electron microscope that forms the first intermediate image. The specimen and objective aperture are typically placed between the upper and lower polepieces of the objective lens.

objective markings

The specifications engraved on the barrel of a light microscope objective. Typically these markings include: manufacturer; objective name; magnification; numerical aperture; immersion medium; application; lens to image distance; coverslip thickness; working distance; code number; and a color-coded ring identifying magnification.

objective prism

The Wollaston or Nomarski prism located in a slot above the objective nosepiece of a DIC microscope. The objective prism can be translated to regulate the amount of bias retardation.

objective stigmator see stigmator

objective turret

A rotatable turret that carries the objectives on a light microscope.

object plane

The plane that contains the object of a lens.

object point

The point on the optical axis intersected by the object plane.

Objective turret or nosepiece of light microscope. Courtesy of Carl Zeiss Ltd.

object space

The space on that side of a lens where the object is located.

observation tube see eyepiece tube

octopole lens

An electron lens with eight magnetic coils arranged in a circle around the optic axis, typically used for aberration correction in electron microscopes and spectrometers.

octupole lens see octopole lens
ocular see eyepiece

off-axis aberrations

Aberrations of lenses that occur with object points that are off the optical axis, i.e. astigmatism, coma, and field curvature.

off-axis electron holography

A technique for the generation of a hologram of a specimen in a transmission electron microscope. In off-axis electron holography, the specimen is placed off the optic axis under illumination with a coherent electron beam. After leaving the objective lens the beam is spit by an electron biprism into two beams, one that has passed through the specimen and a reference beam, which are then recombined to form a hologram.

off-axis holography

The production of a hologram by interference of reference and object beams with different orientations to the specimen. In off-axis holography a coherent beam is divided by a beamsplitter into an object beam which passes through the specimen and a reference beam which passes by the specimen (hence off-axis); the two beams meet at an angle and interfere to form the hologram.

offset

A positive or negative voltage applied to a video signal so that the least intense image points are just detected.

oil-diffusion pump

A diffusion pump that uses oil vapor as the trapping material.

oil-immersion condenser

A light microscope condenser designed for use with an immersion medium between the front lens

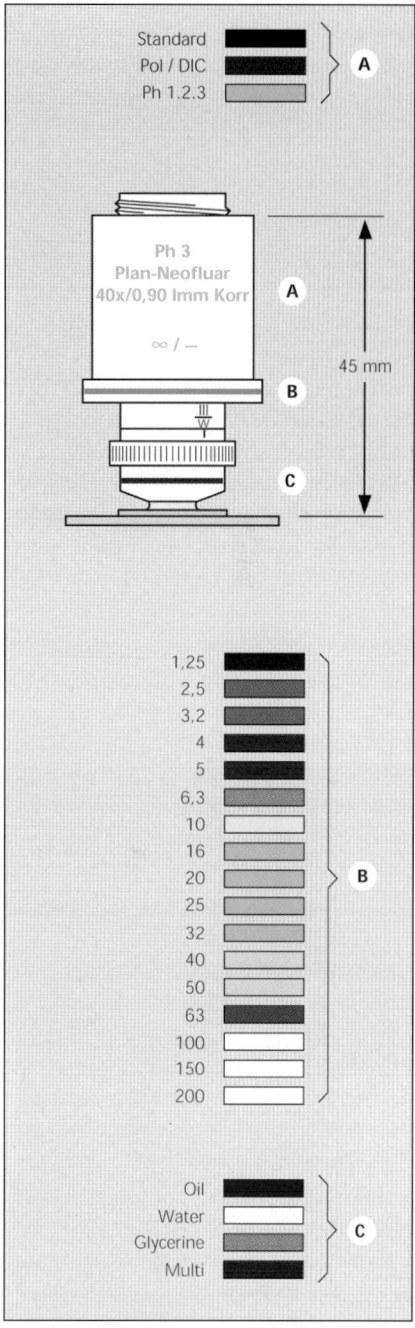

Key to the markings on the barrel of light microscope objectives. Courtesy of Carl Zeiss Ltd.

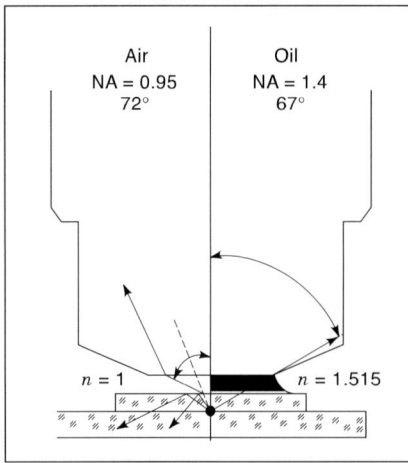

Effect of immersion oil on numerical aperture. From D. Murphy © John Wiley and Sons, Inc.

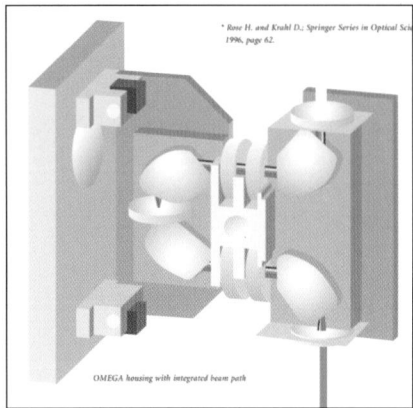

Schematic of omega energy filter. Courtesy of Carl Zeiss SMT.

and a glass specimen support slide.

oil-immersion objective

A light microscope objective designed for use with immersion oil between the front lens and a glass coverslip. The use of immersion oil increases the angular range of diffracted rays captured by the lens, increasing numerical aperture from ~0.95 to over 1.4.

oil-mist filter

A device that traps oil discharged from a rotary pump.

oil tank

A tank containing the transformer and associated circuitry for the high-voltage supply of an electron microscope. The tank contains an insulating oil or gas, such as SF_6.

omega filter

An in-column electron spectrometer used for spectroscopy and energy-filtered imaging in a transmission electron microscope. The omega filter has four 90° magnetic sector prisms, arranged like the Greek letter omega, that deflect the electron beam through 360° and return it to the optic axis. An omega filter is positioned in the microscope column between two of the intermediate lenses. A slit exit aperture allows selection of the energy range that passes on to a detector.

on-axis aberrations

Aberrations of lenses that occur with object points that are on the optical axis, e.g. chromatic and spherical aberrations.

one-ångström microscope

A high resolution transmission electron microscope capable of forming images with a resolution better than one ångström (0.1 nm).

open-loop scanning

The use of a scanning probe microscope without monitoring and correction of the non-linear behavior of the scanner.

optic
1. A component of an optical system, such as an eyepiece or lens.
2. Relating to the eye or the optical properties of a material.

optic axis
1. The axis of an optically anisotropic material about which its atoms or structures are arranged symmetrically. Light passing along the optic axis is transmitted without double refraction; light perpendicular to the optic axis is double refracted.
2. A commonly used synonym for optical axis.

optical activity
A property of a medium in which the plane of vibration of a linearly polarized wave is rotated during transmission.

optical anisotropy
The unequal spatial distribution of optical properties caused by asymmetric atomic or molecular structure. Optical anisotropy is a property of many crystalline materials and is shown as dichroism, pleochroism and double refraction.

optical axis
The axis of symmetry of a lens or a perfectly aligned set of lenses. The optical (or optic) axis is described by the central ray in an optical system, about which the refractive index (in light optics) and fields or potentials (in electron optics) are expanded and from which off-axis distances and angles are measured. In deflectors and prisms, the optical axis is bent or curved.

optical beam-induced current
The current induced in a semiconductor as a result of incident light from a focused laser. By measuring the induced current at each position of a scanning beam, an optical beam-induced current image can be created.

optical bench
A stable frame for the attachment and alignment of lenses and other optical devices.

optical breakdown

The destruction of materials by intense beams of light, typically from lasers.

optical cement

An epoxy adhesive used for bonding optical components, e.g. lenses.

optical coherence tomography

An imaging technique that uses low coherence interferometry to image structures in the superficial layers (<2 mm) of living tissues such as retina and skin. Optical coherence tomography generates images by using laser light (650-1300 nm) and a Michelson interferometer to compare the echo time delay of light that is backscattered from internal structures in tissue with a reference beam.

optical coupler

Any conduit that connects two parts of the optical train of a microscope.

optical density

A measure of the absorption of light. Optical density is the logarithm (to base 10) of the reciprocal of the internal transmittance.

optical diffraction pattern

A diffraction pattern of an object or image illuminated with light.

optical filter

An optical device that changes the intensity, wavelength or state of polarization of transmitted or reflected light.

optical flat

A calibration standard made of optically pure glass with polished plane parallel sides. Optical flats may be used for the calibration of light microscope objectives, such as the measurement of numerical aperture.

optical lever

A cantilever whose movement is detected with a laser beam.

optical microscope
A synonym for light microscope.

optical microscopy
A synonym for light microscopy.

optical orientation imaging
The use of polarized light microscopy for orientation imaging.

optical path difference see optical pathlength difference

optical pathlength
The number of vibrations of a wave passing between two points. The optical pathlength is the product of the refractive index n of the transmitting medium and the distance t traveled. OPL = nt.

optical pathlength difference
The difference in optical pathlength of two waves passing through media of different refractive index and/or thickness. Symbol Δ.

Optical pathlength difference is given by:
$$\Delta = (n_2 - n_1)t$$
where n_1 and n_2 are refractive indices of media 1 and 2, and t is distance traveled.

optical profiler
A profilometer that uses visible radiation.

optical profilometer
A profilometer that uses the interference of light to measure surface topography.

optical scanning probe microscope see nearfield scanning optical microscope

optical section
The image formed at any plane of focus in a specimen.

optical spectrometer
A spectrometer for the analysis of ultraviolet, visible and infrared radiation. An optical spectrometer typically contains a rotatable crystal, diffraction grating or prism that disperses incident

light allowing analysis of each wavelength.

optical transfer function

A measure of the relationship between the amplitude and phase of an object and those of its image. The OTF is a complex function in which the real term is the modulation transfer function and the imaginary term is a phase transfer function.

optical trapping

The spatial confinement of a transparent particle due to the radiation pressure of an intense beam of light. Optical trapping can be explained by the principle of the conservation of momentum: when a photon is refracted away from the beam it loses momentum which is transferred to the refracting object, moving it towards the center of the beam.

optical tubelength

The distance between the back focal plane of the objective and the intermediate image plane of a light microscope.

optical tweezers

The manipulation of particles that are optically trapped using a light microscope with laser illumination. Optical tweezers can be used to investigate the interactions of macromolecules in living cells, colloids and polymers.

optics

The study of the properties of light and optical systems.

O ray see ordinary ray

ordinary ray

One of the two linearly polarized rays (the E and O rays), with mutually perpendicular vibrational planes, transmitted by a birefringent material. The O ray obeys the normal law of refraction.

orientation imaging

The use of polarized light microscopy or electron backscattered diffraction to image crystalline materials and analyze their properties.

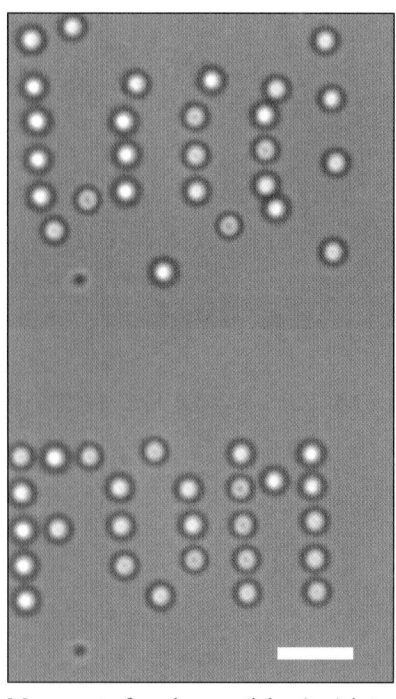

Movement of random particles (top) into a pattern (below) with optical tweezers. Scale bar = 5 μm. *Microscopy and Analysis* 2005.

orientation imaging microscopy

The automated analysis of grain and phase size, orientation and boundaries in polycrystalline materials using electron backscattered diffraction.

orientation mapping

The mapping in a scanning electron microscope of the crystalline microstructure of a specimen by stepping the electron beam across the surface of the specimen and collecting an electron backscatter pattern at each point.

O-ring

A standard type of vacuum seal, comprising a circle of rubber, metal or synthetic material, which provides a demountable air-tight seal for electron microscopes and other vacuum instruments.

orthoscopic mode

The use of a light microscope to view an image of the specimen, the normal mode in a light microscope.

osmium plasma coater

An apparatus for the coating of specimens with osmium by plasma sputtering. Osmium is one of the densest elements and produces a coating that has a fine grain size and high Z contrast in the scanning electron microscope.

osmium tetroxide

A chemical fixative suitable for all biological specimens. Osmium tetroxide crosslinks fatty-acid groups of lipids, adds heavy-metal contrast to weakly scattering biological specimens, and acts as a mordant. It is commonly used as a post-fixative after glutaraldehyde primary fixation.

outgassing see degassing

output node

The location or pixel on a charge-coupled device where charges are converted to a voltage and then exported to the electronic image processing chain.

overfocus

The formation of an image in a plane in front of (upstream of) the true or Gaussian image plane.

overvoltage ratio

The ratio of the applied energy to the critical ionization energy of an electron; symbol U. In X-ray microanalysis the overvoltage ratio determines the accelerating voltage of an electron beam required for maximum emission from any atom.

oyster grid see **folding grid**

P

PAL video standard

The phase-alternating line system, 625-line, 25 frames per second, color broadcast video standard used in Europe.

pancake lens see non-conventional magnetic lenses

paper processor

A machine that develops and dries sheets of exposed photographic paper.

parabolic mirror

An aspherical, paraboloidal mirror widely used in optical devices to produce a beam of parallel light from a source placed at its focus.

paraboloid condenser

A condenser used for darkfield light microscopy. A paraboloid condenser has an annular diaphragm that transmits light onto a circumferential paraboloidal mirror from which light is reflected at an oblique angle onto the specimen.

paraffin wax

An embedding medium widely used in histology. Melting point: ~55°C.

paraformaldehyde see formaldehyde

parallax

The apparent lateral movement of an object with respect to others when viewed from two different directions, e.g. by each eye or by each eyepiece in a stereomicroscope. Parallax allows depth perception and stereoscopic vision.

parallel-beam X-ray spectrometer

An X-ray spectrometer with a focusing optic placed close to the specimen that produces a parallel beam of X-rays entering the spectrometer.

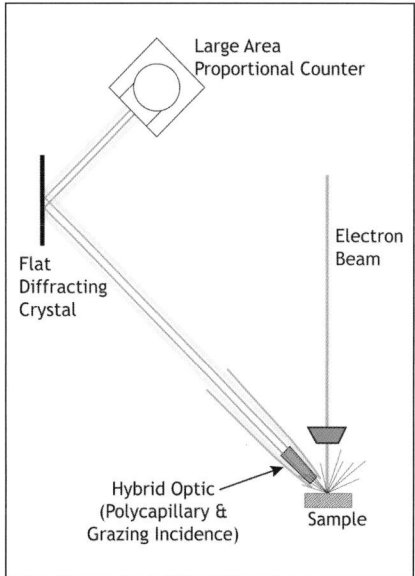

Principle of parallel-beam X-ray spectroscopy. Courtesy of Thermo Electron Corporation.

parallel electron energy-loss spectroscopy

Electron energy-loss spectroscopy in which each channel of an EELS spectrum is acquired simultaneously with a parallel electron energy-loss spectrometer.

parallel energy-loss electron spectrometer

An electron spectrometer designed for the simultaneous acquisition of all channels of an electron energy-loss spectrum. The spectrometer is equipped with a post-prism set of magnetic lenses that focus the dispersed electrons onto a one- or two-dimensional array of photodiodes that produce the energy-loss spectrum or an energy-filtered image.

parallel energy-loss electron spectrometer detector

A photodiode detector used in parallel energy-loss electron spectroscopy. A PEELS detector typically comprises a YAG scintillator coupled by fiber optics to a one-dimension array of 1024 25-mm diameter photodiodes.

parallel register

The rows of pixels in a CCD that accumulate or store charges and transfer them in parallel to the serial register.

parasitic aberrations

Aberrations of lenses that are the consequence of mechanical or similar imperfections in lens construction, particularly in electron lenses. For round magnetic lenses, the dominant parasitic aberrations are astigmatism and coma.

paraxial approximation

A method used in first-order or paraxial optics that assumes that for small angles to the optical axis sines and tangents of angles are equal to the angles. The approximation holds for ray angles of up to 30° but using the paraxial approximation for larger angles can lead to aberrations.

paraxial image plane

The position of the image plane as calculated using paraxial optics.

paraxial optics

A method in optics that assumes that all ray angles are small and that all rays travel on or close to the optical axis.

paraxial ray

A ray that passes along or close to the optical axis of a lens or imaging system.

parfocal

A term used to describe a set of objectives in the turret of a light microscope that can be interchanged without changing focus, a standard design in light microscopes. Parfocal objectives have identical distances from the specimen to their insertion point in the objective turret.

partial-pressure gauge

A vacuum gauge that measures the partial pressure of one molecular species, typically by mass spectrometry.

particle analysis

A procedure in image analysis for the quantification of the number and parameters of particulate objects in an image.

particle wave

The wave produced by the interference of diffracted and surround waves in the intermediate image plane of a microscope.

pascal

SI derived unit of pressure. Symbol Pa. One pascal equals one newton per square meter. 1 pascal = 0.0075006 torr, or 0.01 mbar.

passband see bandpass

patch stop

A central opaque disk that blocks direct propagation of a beam, commonly used for darkfield microscopy.

peak overlap

The overlap of characteristic peaks in an X-ray microanalysis spectrum due to similarity of X-ray

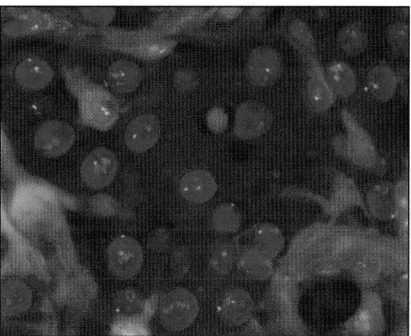
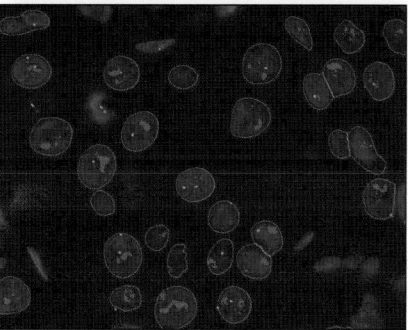

Particle analysis. Top: Fluorescent in-situ hybridization of breast carcinoma tissue showing FITC-labeled PNA of the centromer region of chromosome 17 (green dots), Texas Red-labeled DNA of the HER2/neu gene (red dots) and nuclei counterstained with DAPI (blue). Bottom: With particle analysis software, the colocalization of signals can be demonstrated and quantified. Courtesy of Soft Imaging System.

emission energy and/or the energy-resolution limitation of the detector. Peak overlap can prevent the unambiguous detection of the presence (or absence) of a specific element.

peak-to-background ratio

The ratio of the height of a characteristic peak to background bremsstrahlung in an X-ray microanalysis spectrum.

peak wavelength

The wavelength that is maximally emitted or transmitted.

Peltier cooling

The cooling of a camera, detector or instrument by the Peltier effect.

Peltier effect

The drop in temperature when current flows between two different metals or semiconductors. The temperature change is directional; reversing the current causes heating. Temperatures 50°C below ambient are typically achievable.

Penning gauge

A cold-cathode pressure gauge that measures the electrical current produced by the ionization of molecules. In a Penning gauge a high voltage (2-5 kV) electric field between the cathode and anode ionizes gases; positive ions are attracted to the cathode and electrons to the anode creating a current. The gauge measures gas pressures in the range 10^{-2} to 10^{-7} Pa.

perfusion fixation

The chemical fixation of a living organism, organ or tissue by perfusion of fixative through the vasculature.

petrographic microscope

A polarized light microscope designed for the study of rocks and minerals.

Petzval field curvature see field curvature

phase

1. The position of a cyclical, periodic or wave

motion, typically in relation to that of another. 2. One physically distinct component of a heterogeneous system, such as ice and water.

phase-alternating line system see PAL video standard

phase annulus

A synonymous term for the annular diaphragm placed in the front focal plane of the condenser in a phase contrast light microscope.

Phase annuli and phase contrast condenser. Courtesy of Carl Zeiss Ltd.

phase contrast

The generation of contrast by conversion of differences in the phase of waves emerging from an object, detector or probe into amplitude differences in the image.

phase contrast, in SPM

The generation of image contrast in a scanning probe microscope by comparing the phase differences between the electrical signal that generates the motion or resonance of a cantilever or probe with the signals that are collected from the probe. The slight differences, or time delay, in these signals yields information about the interaction of the probe and specimen such as elasticity, viscosity and adhesion.

phase contrast, in TEM

The generation of image contrast by interference of electron waves. Many imaging modes in TEM and HREM generate phase contrast whenever the image is formed by more than one beam.

phase-contrast light microscope

A light microscope with accessories for phase contrast imaging. The accessories are an annular diaphragm in the front focal plane of the condenser and a phase contrast objective. The condenser projects an image of the annulus at infinity, illuminating the specimen with parallel light; the objective forms an image of the condenser annulus at its back focal plane, thus condenser annulus and phase plate are conjugate. To align the microscope for phase contrast microscopy the condenser annulus must be centered on the annulus of the phase plate using a Bertrand lens or phase

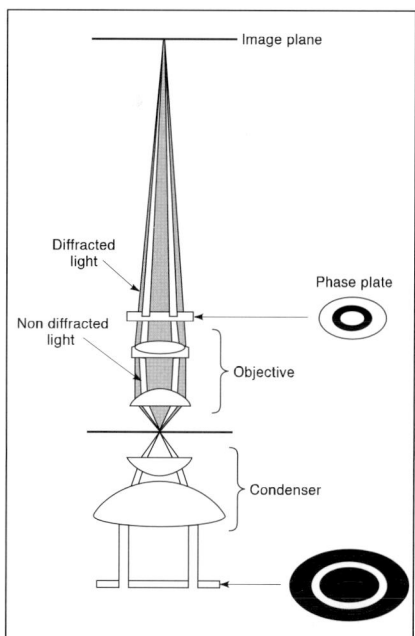

Principle of image formation in a phase-contrast light microscope. From D. Murphy © John Wiley and Sons, Inc.

239

telescope. Light that is not diffracted by the specimen (the S wave) falls on the annulus or ring in the phase plate, whereas light that is diffracted by the specimen (the D wave) experiences a typical phase retardation of up to λ/4 and is focused across the complementary area of the phase plate. The phase plate adds a further relative phase shift of λ/4 between S and D waves which leave the plate with a total relative phase shift of up to λ/2. The S and D waves destructively interfere when recombined in the intermediate image plane giving rise to amplitude contrast in the image that is directly related to the phase shifts caused by local changes in specimen refractive index. In positive phase contrast microscopy phase objects appear dark on a gray background; in negative phase contrast microscopy phase objects appear bright on a darker background.

phase-contrast light microscopy

The use of a phase-contrast light microscope.

phase-contrast objective

A type of light microscope objective with a phase plate in its back focal plane for phase contrast microscopy. Phase contrast objectives are marked with a code on the barrel to allow selection of the appropriate condenser phase annulus.

phase halo

The bright zone at the edge of objects when viewed with phase contrast light microscopy. Phase halos occur when the image of the condenser annular diaphragm in the back focal plane of the objective and the phase plate annulus have different dimensions, allowing some of the light diffracted from objects with low spatial frequencies to pass through the phase plate with the surround wave and emerge without net retardation. The phase halo can be reduced by increasing the refractive index of the object immersion medium or by the use of apodized phase contrast objectives

phase imaging mode, in SPM see phase contrast, in SPM

phase mask

A diffraction grating that acts as a beamsplitter for incident monochromatic light. A phase mask typically consists of a surface relief grating etched on fused silica.

phase object

An object or specimen that changes the phase of a transmitted wave.

phase plate, in light microscopy

A spatial filter used in phase contrast objectives. A phase plate has a central annulus or ring and shifts the phase of light transmitted by the ring by one quarter wavelength relative to that passing through the complementary area of the plate. In older systems the phase shift is introduced by making the ring either thinner or thicker than the rest of the plate; in newer systems the same effect is achieved with phase-retarding dielectric layers in or around the ring. The annulus is coated with a metal film to attenuate surround light by 60-90%.

phase plate, in TEM

A thin film of electron transparent material, such as carbon, that alters the phase of a transmitted electron beam.

phase ring see phase annulus, phase plate

phase shift

The change in phase of a wave relative to background waves. Symbol δ. The phase shift in radians is equal to $2\pi\Delta/\lambda$ where Δ is optical path difference and λ is wavelength.

phase telescope see centering telescope

phonon

A quantum of energy associated with the collective vibration of the atoms in a crystal lattice.

phosphor

A substance that exhibits phosphorescence. Phosphors such as CdS and ZnS are used as coatings in detectors and on viewing screens.

Phase masks. Courtesy of StockerYale Inc.

phosphorescence

Luminescence that occurs on a timescale from 10 ns to several seconds. Phosphorescence involves transition of an excited molecule through a triplet energy state before relaxation to ground state and release of a photon.

phosphor screen

A screen coated with a phosphor that displays an image. Phosphor screens are used in cathode ray tubes and transmission electron microscopes.

photobleaching

The irreversible reduction in the emission of a fluorophore caused by damage to the molecule by physical or chemical action. Photobleaching is typically performed with a short pulse from a high-power laser.

photocathode

A cathode that emits electrons by the photoelectric effect. A photocathode may be a metal plate coated with a photoluminescent material or a semiconductor. Photocathodes are used as primary transducers in many imaging detectors, e.g. photomultiplier tubes.

photodetector

A general term for a detector of photons.

photodiffraction

The diffraction of light.

photodiode

A semiconductor-based light detector. Photodiodes typically consist of a reverse-biased p-n junction; this design produces a high resistance which is reduced when photons excite the semiconductor allowing current to flow in proportion to the incident light.

photodiode array see charge-coupled device

photoelasticity

The development of optical anisotropy as a result of applied compression or tension.

photoelectric effect
The emission or liberation of electrons as a result of excitation by photons.

photoelectron
An electron emitted from an atom as a result of excitation by a photon. Photoelectrons are used to analyze specimens in X-ray photoelectron spectroscopy.

photoelectron microscopy see X-ray photoemission microscopy

photoemission electron microscopy
The use of photoelectrons, emitted from the specimen by excitation with UV or visible light sources, for imaging and spectroscopy.

photo eyepiece see projection eyepiece

photographic enlarger
An optical device that projects an image of a film negative onto photographic paper for the production of photographs.

photography
The recording of an image on film or a detector.

photoluminescence
Luminescence produced by visible light.

photomacrography
The production of a photographic image with no or low magnification of the object on the film.

photometric contrast
The ratio of the intensities of objects or of an object to its background in an image.

photomicrograph see micrograph

photomicrography
The recording on film or by a detector of an image produced by a microscope.

photomultiplier
A detector that amplifies photon signals. In a

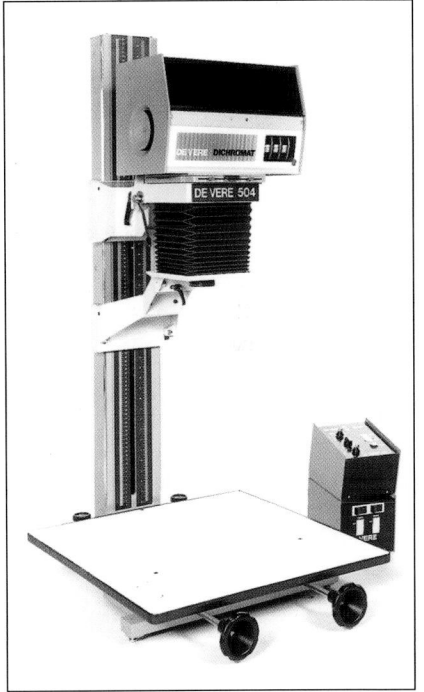

Photographic enlarger. Courtesy of Agar Scientific.

photomultiplier, photons strike a photocathode generating electrons that are accelerated through a vacuum tube and increase in number by serial collisions with, and secondary emission from, a chain of dynodes. The final amplified signal, which is directly proportional to the input, is transmitted by an anode or collection electrode.

photon

An elementary particle or quantum of electromagnetic radiation. Photons travel at the speed of light. The energy in electronvolts of a photon is equal to hf, where h is Planck's constant and f is the frequency in Hz.

photon noise see shot noise
photon shot noise see shot noise

photonic force microscope

A microscope that uses an optically trapped particle as a probe to map or measure the topography or physical properties of a specimen by detection of the displacement of the trapped particle.

photonics

The study of photons and the technologies and applications that use them.

photothermal microscopy

The use of a light microscope to detect local temperature changes in living cells and tissues. In photothermal microscopy the specimen is irradiated for several ns with a pulsed laser beam, inducing local heating. A second laser pulse is then used to detect changes in the refractive index of the sample.

phototoxicity

Any damage to a specimen caused by the illumination source while under observation on a microscope. Phototoxicity may be caused by physical (e.g. heating) or chemical (e.g. free radical) interactions.

phototube

A part of the tube of a light microscope that

carries an eyepiece designed for use with a
detector such as a camera.

physiological contrast

The ability of the human visual system (eye and
brain) to discriminate detail in an image formed on
the retina. If differences between details are due to
brightness, the contrast is called brightness
contrast; if they are due to chromaticity, it is called
color contrast.

picture element see pixel

piezoceramic

A ceramic material that demonstrates the piezoelectric effect, such as lead zirconium titanate
(PZT).

piezoelectric actuators

A device that uses the piezoelectric effect to
transform an electrical input into movement.

piezoelectric crystal

A crystal that demonstrates the piezoelectric effect.
In a piezoelectric crystal, electron-hole pairs are
separated but symmetrically distributed.
Application of stress such as bending produces
charge asymmetry and generates a voltage. The list
of piezoelectric crystalline materials is long and
includes crystal quartz, Rochelle salt, bone, and
certain ceramics, metals and polymers.

piezoelectric effect

The generation of charge asymmetry in certain
materials in response to mechanical stress. The
effect is reversible: an applied voltage can cause a
change of shape, a phenomenon called the
converse or inverse piezoelectric effect. A typical
response is on the order of nanometers per volt.

piezoelectric scanner

A scanner that uses the the piezoelectric effect to
move a specimen or probe.

piezoforce microscopy see piezoresponse force microscopy

piezolever see piezoresistive cantilever

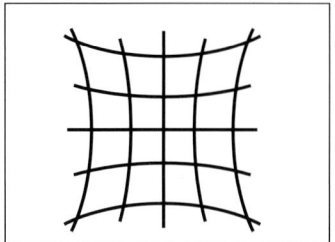

Pincushion distortion. From D. Murphy © John Wiley and Sons, Inc.

piezoresistive cantilever

A type of cantilever manufactured from or containing a piezoresistive material, such as doped silicon, that changes its resistance upon deformation. Piezoresistive cantilevers do not require a laser-based motion detecting device as the cantilever can itself report deflection.

piezoresponse force microscopy

The detection and analysis of the mechanical response of piezoelectric and ferroelectric materials to a voltage bias applied by the tip of a scanning probe microscope.

piezoresponse imaging see piezoresponse force microscopy

pincushion distortion

A type of distortion where the image shows increasing magnification from center to margins; a square object will appear pincushion shaped.

pinhole aperture

A variable diaphragm in front of the detector in a confocal microscope that regulates spatial resolution and signal to noise ratio in the image. Typical pinhole diameters are around 100 μm.

Pirani gauge

A pressure gauge that measures the rate of dissipation of heat in the reference space. In a Pirani gauge, the currents required to maintain a constant temperature of two heated wires, the reference wire in a sealed vacuum tube and the detector wire in the reference space, are compared: any gas molecules will produce convective cooling of the detector wire, increasing the heating current. The gauge measures gas pressures in the order of 10^{-1} Pa.

Condenser lens

Beam shift and tilt coils

Plane of pivot point

Image plane

Pivot point for oscillation of electron beam between two tilt conditions. *Microscopy and Analysis* 2004.

pivot point

The point on the optic axis of an electron microscope about which the beam pivots when tilted. The position of the pivot point is a function of the ratio of the excitation of the upper and lower deflection coils that produced the tilt.

pixel

An abbreviation for picture element. A pixel is the smallest discrete value of a digital image or the smallest element of a detector or monitor.

pixelation

1. A synonym for the digitization of an image.
2. The result of having a small number of pixels in a digital image, so that the pixels are clearly seen by the observer.
3. Empty magnification of a digital image.

pixel shift

The movement of an image point when changing accessory optical elements. Pixel shift occurs in fluorescence microscopy when changing barrier filters to view specimens labeled with differently colored fluors.

pixel shifting

A technique for increasing the number of pixels in an image formed by a CCD, allowing the resolution of specimen detail smaller than one detector pixel. In pixel shifting the CCD is physically displaced with a piezo crystal a subpixel distance (usually one half) horizontally and/or vertically between several image captures. Software is then used to combine the images and output a single image with many more pixels than the CCD.

Planapochromatic objective lenses. Courtesy of Nikon.

planapochromat

An apochromatic lens or light microscope objective with a flat field of view.

planapochromatic objective see planapochromat

planar grinder

A grinder used for the preparation of parallel-sided specimens for further processing for transmission electron microscopy.

Planck's constant

Symbol h. A constant relating a quantum of energy E to its frequency v: $E = hv$. The Planck constant is 6.626×10^{-34} joule second.

Planar grinder. Courtesy of E.A. Fischione Instruments, Inc.

plane of focus see **focal plane**

plane of vibration
The plane parallel to and passing through the axis of propagation of an electromagnetic wave, containing the electric field vector and the propagation vector.

plane-polarized light see **linearly polarized light**

planoconcave lens
A diverging lens with one concave and one planar surface.

planoconvex lens
A converging lens with one convex surface and one planar surface.

PlasDIC
A brand-specific mode of polarized light differential interference contrast light microscopy designed for optically anisotropic specimens such as living cells in plastic culture dishes.

plasma cleaning
The cleaning of specimens by mild etching with a plasma, typically to remove surface contamination such as hydrocarbons.

plasma cleaning device
An apparatus that removes surface contamination from specimens using a plasma. A plasma cleaning device has a vacuum chamber in which a plasma is created from introduced gases, typically oxygen and argon, by an electrical arc or high frequency oscillating electric field. The plasma dislodges or chemically reacts with hydrocarbons on the specimen forming water and CO_2 which are removed by the vacuum pump.

plasma etching
The etching by an ionized plasma.

plasma etching device
A device for the removal (etching) of surface layers of a specimen using a plasma of gas such as

Plasma cleaning device. Courtesy of E.A. Fischione Instruments, Inc.

Ar, N, O or XeF_2. A plasma-etching device operates at a higher wattage than a plasma cleaner, and removes significant amounts of the surface of a specimen, exposing deeper layers for microscopical analysis.

plasmon

A quantum of energy associated with the longitudinal collective oscillations of weakly bound electrons in a solid or plasma.

plasmon peak

A peak in an electron energy-loss spectrum contributed by plasmon excitation.

plastic film

A support film made of plastic material, e.g. collodion, Formvar. Plastic films are usually made by evaporating an organic solvent solution of the plastic on a planar surface such as water or glass. The films can then be picked up on a support or electron microscope grid.

plastic section

A common term for a section of resin-embedded material.

plasticizer

A compound added to a resin to increase flexibility after polymerization.

plate

1. A sheet of photographic film.
2. A printed photograph in a publication. The term plate arises from the original use of glass plates to carry film emulsion.

pleochroism

The variation in color of an optically anisotropic material when illuminated along different axes. Pleochroism is caused by the preferential absorption of specific wavelengths or polarization states of light.

plunge freezer see plunge freezing device

plunge freezing

The cryofixation of a specimen by plunging at

Plunge freezing device. Courtesy of Gatan, Inc.

high speed (>1 m s^{-1}) into a cryogen. Plunge freezing is suitable for small specimens or particles on EM grids.

plunge freezing device

A device for the cryofixation of specimens by plunging them into a cryogen at high speed. Plunge freezing devices operate by gravity, or by using electromagnets to accelerate the specimen. Plunge freezing devices allow rapid freezing of thin film suspensions on EM grids for cryo-TEM.

plural scattering

The scattering of an electron or photon as a result of consecutive interactions with between 2 and 20 atoms.

p-n diode see semiconductor

Pockels cell

A device that changes refractive index due to the Pockels effect.

Pockels effect

The change in refractive index of a transmitting medium as a result of an applied electrical field.

point (to point) resolution

The smallest distance between two clearly resolved points in an image.

point projection microscope

A lensless microscope with a point source of electrons or X-rays that forms a projected (shadow) image of the specimen at a detector. The magnification is the ratio of the distances from specimen to detector and source to specimen.

point source

A source that emits radiation or electrons from a very small region of its surface, e.g. a field-emission filament.

point-spread function

The degree to which the image of a point object is degraded by the optics of a diffraction-limited optical system. In an aberration-free light or electron microscope the point-spread function causes a point object to be imaged as an Airy pattern. A Fourier transform of the point-spread function gives the contrast transfer function.

Poisson noise see shot noise

polarization cross

The dark cross formed in the back focal plane of a polarized light microscope when the objective is fully illuminated and the polarizer and analyzer are crossed.

polarization microscope see polarized light microscope

polarization microscopy see polarized light microscopy

polarization states

The plane(s) of vibration of the electric-field vector of a light wave. Natural light, with random planes of vibration, can be modeled as two mutually orthogonal, incoherent, linearly polarized waves of identical amplitude. By using this model, changes in the properties of the two orthogonal waves can describe linear, circular and elliptical polarization states.

polarized glasses/spectacles

A pair of glasses (spectacles) having polarizers with mutually orthogonal planes of polarization in front of each eye, used for viewing polarized stereoscopic images and displays.

polarized light

Light waves in which the planes of vibration of the electric field vector are confined and not random.

polarized light microscope

A light microscope with accessories for polarized light microscopy. The standard accessories are: a sub-stage linear polarizer; a strain-free condenser; a rotating stage; strain-free objectives in a centerable nosepiece; a compensator or retardation plate; and an analyzer in the tube. The polarizer and analyzer are orientated at right angles, typically east-west and north-south respectively, to

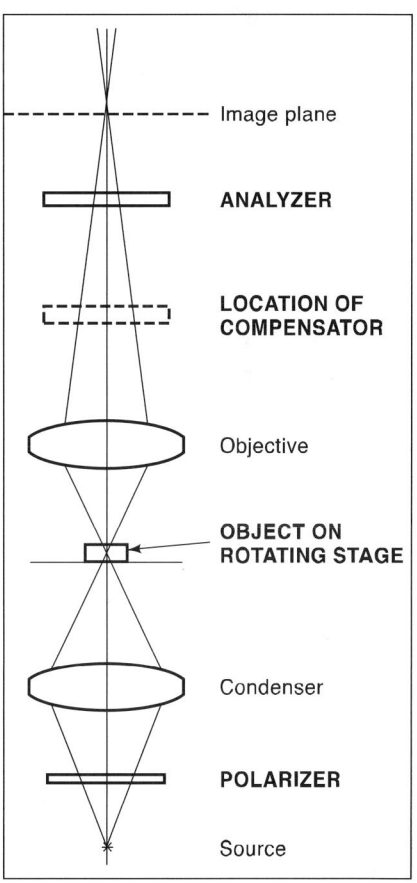

Configuration and optical train of a polarized light microscope. From D. Murphy © John Wiley and Sons, Inc.

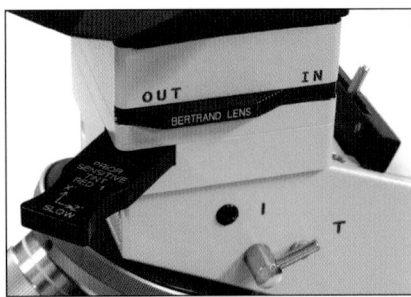

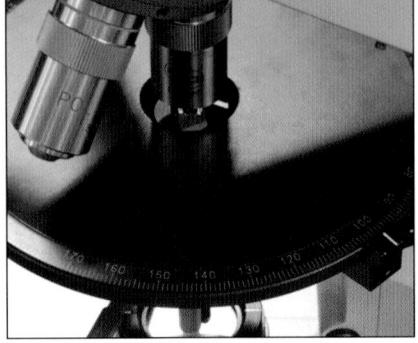

Polarized light microscope showing the ports in the tube for the analyzer and compensators (top) and the rotating stage (bottom). Courtesy of Prior Scientific Instruments.

produce maximum extinction. Any optically anisotropic material placed in this light path will produce ordinary (O) and extraordinary (E) rays with a relative phase shift; the O and E rays are recombined by a compensator that adds an additional phase retardation and recombines the rays to form linearly, elliptically or circularly polarized light that is either transmitted or blocked by the analyzer. The transmitted rays then interfere in the intermediate image plane producing an amplitude image with contrast and colours that are directly related to path differences in the anisotropic specimen. Unknown specimens can be identified with the aid of reference charts such as the Michel-Levy interference color chart.

polarized light microscopy
The use of a polarized light microscope.

polarized stereo display
A stereo display that uses polarized light with orthogonal orientations for each eye and that is viewed with polarized glasses.

polarizer
Any device that changes the state of polarization of light, e.g. a polarizing filter, prism or a reflective surface.

polarizing filter
An optical filter that transmits linearly polarized light. Polarizing filters are typically made from dichroic materials or birefringent crystals.

polarizing prism
A prism that polarizes natural light. A polarizing prism is typically formed by joining two pieces of optically anisotropic material that split incident light into two orthogonally polarized beams, one of which is diverted by internal reflection, allowing a single linearly polarized beam to emerge.

Polaroid
Proprietary name for a type of polarizing filter. Polaroid is made from sheets of orientated polyvinyl alcohol molecules impregnated with iodine. Light waves with planes of vibration parallel to the molecules are absorbed due to

excitation of iodine electrons, whereas waves with other planes of vibration are transmitted.

polepiece
The region of an electromagnet at which the magnetic field is most concentrated. In standard round magnetic lenses used in electron microscopes the polepieces lie inside the bore of the lens.

polishing
The smoothing of a rough surface.

polisher see **rotary polisher**
polycapillary optic see **capillary optic**

polychromatic radiation
Radiation with a broad range of energies, frequencies or wavelengths.

polychromator
An optical spectrometer with multiple exit slits on the Rowland circle allowing the simultaneous analysis of many wavelengths.

polymerization
The formation of stable chemical bonds between molecules. Embedding resins are polymerized by heating, chemical treatment, or ultraviolet light, at high or low temperatures.

polymerizing oven
An oven used for the polymerization of resins at high temperatures.

polyvinyl formal see **Formvar**

polyvinylpyrrolidine
A compound that acts as a plasticizer in embedding media, such as sucrose, used for cryosectioning.

positive eyepiece see **external-diaphragm eyepiece**

positive lens see **converging lens**

Post-column CCD camera on transmission electron microscope. Courtesy of Carl Zeiss SMT.

positive phase contrast

A mode of phase contrast light microscopy where objects with a high refractive index appear dark on a lighter background in the image. Positive phase contrast is the conventional mode in a phase contrast light microscope and uses a positive phase plate in the objective.

positive phase plate

A phase plate with an annulus formed by an absorbing metallic film, surrounded by a complementary area coated with $\lambda/4$ phase-retarding dielectric layer. The absorbing film attenuates the surround waves, reducing background intensity. The phase-retarding layer adds a $\lambda/4$ retardation to the diffracted waves.

positive replica

That surface of a replica with a topography that is an exact copy of that of the specimen.

positive stain

A stain that binds to the components of a specimen giving additional contrast to the specimen.

post-column camera

A camera placed beneath the column of a transmission electron microscope.

post-column energy filter

An electron spectrometer placed beneath the column of a transmission electron microscope.

post-embedding labeling

The application of cytochemical probes to sections of embedded specimens. Post-embedding labeling improves access of probes to the interior of a specimen, but may reduce probe affinity because of changes caused by the embedding procedures.

post-fixation

The secondary or tertiary chemical fixation of a specimen.

powder diffraction

Electron or X-ray diffraction of powdered crystalline materials or specimens containing many small randomly orientated crystals.

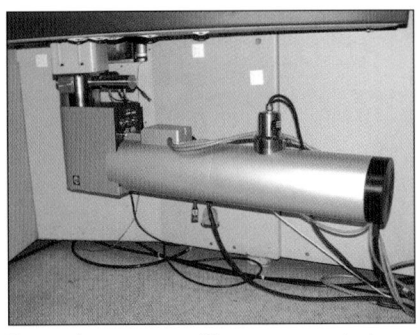

Post-column energy filter on transmission electron microscope. Courtesy of Gatan, Inc.

powder diffraction pattern

A diffraction pattern in which the random orientation of crystal particles gives rise to reflections that are radially arranged around the direct beam and may coalesce to form a ring.

power spectrum

A plot of the range of spatial frequencies in an image, usually demonstrated with an image of the Fourier transform.

p polarized

Radiation polarized parallel to a plane of incidence.

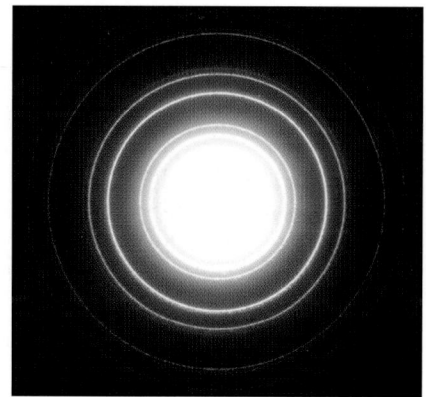

Electron diffraction pattern of randomly orientated evaporated thallous chloride crystals. Courtesy of Agar Scientific.

preamplifier

An amplifier that provides the first amplification of a signal, usually located close to the source of a signal. In an X-ray spectrometer, the preamplifier amplifies the signals from the field-effect transistor and presents them to the pulse processor.

pre-embedding labeling

The application of cytochemical probes to living or fixed specimens before embedding and sectioning. Pre-embedding labeling allows attachment of probes to native or lightly fixed targets, free of embedding matrix, but may reduce probe access to the interior of a specimen or cause trapping of probes.

pressure-limiting aperture

A diaphragm that restricts the diffusion of gases allowing the maintenance of different pressures in compartments on either side of the diaphragm. Pressure-limiting apertures are placed between the column and projection chamber of electron microscopes.

prevacuum pump see rotary pump
primary aberrations see monochromatic aberrations

primary color

Any one of a set of colors that when combined make white. The standard primary colors for images and monitors are red, green and blue.

primary extinction

The loss of intensity in a diffraction pattern as a result of multiple reflections of beams between lattice planes leading to destructive interference with the primary reflection.

primary image

The first image formed by an optical system. The primary image is called the intermediate image if it forms the object of additional optics.

primary image plane

The plane of the primary image formed by an optical system.

principal focal point

The point on the optical axis where rays are focused.

principal focus see principal focal point

principal plane

The plane perpendicular to the optical axis of a lens from which focal length and object or image distances are measured. An infinitely thin lens has one principal plane bisecting the lens; a thick lens has two principal planes, one on the object side and one on the image side.

principal points

The points on the optical axis intersected by the principal planes.

principal ray

A ray that passes through the principle points.

principle of reciprocity

A principle of light and electron imaging systems which asserts that the system will behave in an identical manner if the direction of radiation is reversed. Hence if the object in a light microscope were to be placed in the intermediate image plane, under reversed illumination a demagnified image would appear in the object plane. Similarly, a TEM is reciprocal to a STEM, since ray paths linking points in the electron source and image planes of a TEM can be reversed to reproduce the ray paths in a STEM.

Reciprocity also applies in scanning probe microscopy: by reversing object and image the structure of the probing tip may be examined.

prism

A device that disperses or changes the direction of radiation. Glass polyhedral prisms are used to disperse light, change its state of polarization, or change the orientation of an image. A magnetic prism disperses or changes the paths of electrons.

probe

1. A focused beam of radiation.
2. A scanning probe microscope tip.
3. A chemical, element or molecule used in cyto-chemistry or histochemistry.
4. A finely pointed tool.
5. Anything used to investigate a specimen.

probe current see beam current

profilometer

An instrument that measures surface roughness. Profilometers are of two types, contact and non-contact, according to whether or not there is direct contact of the profilometer probe with the specimen.

profilometry

The use of a profilometer to obtain topographical information.

programmable-array microscope

A fluorescence light microscope that uses a spatial light modulator to define patterns of excitation light or detected emission.

progressive-scan charge-coupled device

A type of charged-coupled device that forms a full frame image one line at a time, without interlacing of video fields.

projection chamber

The chamber at the base of the column of a transmission electron microscope into which the image of the specimen is projected onto a viewing screen, film or detector.

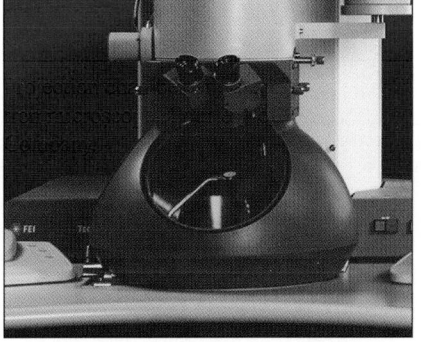

Projection chamber of transmission electron microscope. Courtesy of FEI Company.

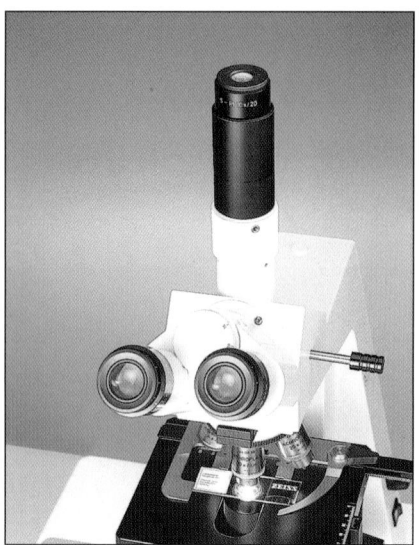

Projection eyepeice on viewing head of light microscope. Courtesy of Carl Zeiss Ltd.

projection eyepiece

An optical device with a projection lens placed in the eyepiece tube of a light microscope to project an image into a camera.

projection lens

A lens that forms a real image at a finite distance.

projection microscope see microprojector, point projection microscope

projector coils

A set of double-deflection coils, located above the projector lens(es), that control the alignment of the image in the electron microscope column. The console controls are usually labeled image tilt and image shift.

projector lens

The final lens in an electron microscope that magnifies the last intermediate image and projects it onto the fluorescent screen. The penultimate lens may also be called a projector lens.

propane-jet freezer see propane-jet freezing device

propane-jet freezing

A method of cryofixation in which the specimen is frozen by a high-pressure jet of liquid propane.

Propane-jet freezing device. Courtesy of Boeckeler Instruments, Inc.

propane-jet freezing device

A device for the cryofixation of specimens by jets of liquid propane. A propane-jet freezing device has two nozzles that propel high-pressure jets of liquid propane (~-180°C) onto two copper plates that sandwich the specimen.

protease digestion see antigen retrieval

protein crystallography

The study of the atomic structure and 3D conformation of crystalline proteins by electron and X-ray diffraction.

ptychography

The extraction of phase information from the

interference patterns generated by overlapping disks in a diffraction pattern obtained using a scanning transmission electron microscope.

pulse processor

An amplifier used in an X-ray spectrometer that converts voltage steps from the field-effect transistor into separate pulses with amplitudes proportional to X-ray energy.

pulsed force microscopy

Atomic force microscopy in which the specimen stage (z piezo) is oscillated at 0.1-2 kHz with an amplitude of 10-500 nm, so the specimen cyclically interacts with a resonating probe in order to generate force-distance curves. PFM reveals specimen topography, adhesion and stiffness.

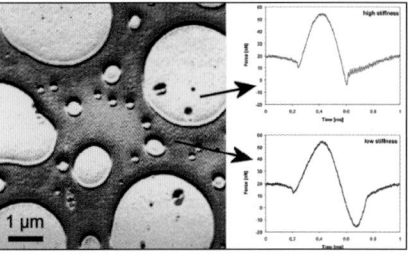

Pulsed force microscopy. Stiffness image of a thin film of polymethylmethacrylate containing styrol-butadiene rubber, with corresponding PFM curves. Courtesy of WITec.

pupil

An image of the aperture stop of an imaging system. A pupil defines the cones of light entering and exiting an optical system from any object point.

pupil, of the eye

The variable aperture of the iris diaphragm of the eye. The diameter of the pupil varies from 2 mm in bright light to 8 mm in the dark.

P wave see **particle wave**

Q

quadrant photodiode

A photodiode detector with four light sensing regions, allowing the output signals to be compared.

quadrupole lens

An electron lens comprising four magnetic coils or four electrodes arranged in a circle around the optic axis, typically used for aberration correction in electron microscopes and spectrometers. In the plane containing two of the electrodes or midway between two poles, the lens acts as a convergent lens; in the plane normal to this, it acts as a divergent lens.

qualitative microanalysis

The use of microanalysis to determine the presence and spatial distribution of elements in a specimen.

quality factor

A measure of the damping of an oscillating object. Symbol Q. The quality factor of a cantilever in a scanning probe microscope is a measure of force sensitivity and is given by:
$$Q = (f_o m)/b$$
where f_o is free resonant frequency, m is mass and b is a damping factor.

quality factor control

The active regulation of the quality factor of a cantilever in response to tip-specimen interactions by means of a positive feedback loop in the cantilever vibration circuit.

Quantifoil

A brand-specific carbon support film with an array of circular or square holes.

quantitative microanalysis

The use of microanalysis to quantify the elements in a specimen.

quantity of light see **luminous energy**

quantum dots

Nanometer-sized semiconductors used as probes in fluorescence microscopy and electron microscopy. Quantum dots are 2-20 nm wide nanocrystals of materials such as CdSe and InGaP and display an emission that is composition- and size-dependent, tuneable and resistant to photobleaching. Emission is regulated by quantum confinement, so as dot size increases, emission shifts to longer wavelengths for the same excitation wavelength.

quantum efficiency see detector quantum efficiency

quantum yield

The ratio of the number of excitation or emission events to the number of photons absorbed by a system. Symbol ϕ.

quarter-wave plate

A retardation plate that produces an optical path difference or retardation of one quarter wavelength. Linearly polarized light with a plane of vibration at an angle of 45° to the optic axis is transmitted as circularly polarized light, and vice versa.

quartz

The common name for silica or silicon dioxide (SiO_2). Fused quartz is the optically isotropic form of quartz used in the manufacture of lenses (RI at 500 nm: 1.462). Crystal quartz is the optically anisotropic, piezoelectric form of quartz and is used to make semiconductors and polarizing devices. (RI at 532 nm: n_o 1.547; n_e 1.556).

quartz-crystal monitor

A film thickness meter used in coating procedures. The meter monitors the change in resonant frequency of an oscillating quartz crystal as its mass increases due to film material condensing on its surface. The monitor can measure film thicknesses to 0.1 nm.

quartz halogen lamp

An incandescent lamp with a tungsten filament

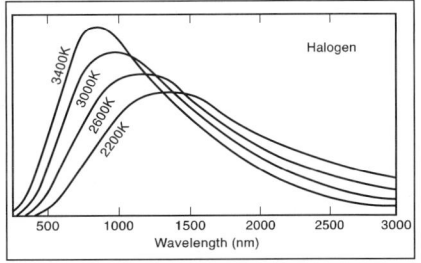

Emission spectrum of quartz halogen lamp. From D. Murphy © John Wiley and Sons, Inc.

and a quartz bulb filled with a halogen gas, typically bromine or iodine. The halogen interacts with evaporated W atoms, reducing any coating of the inside of the bulb; when the halogen-W complexes impact the filament they dissociate, depositing W atoms back on the filament. This allows the filament to operate at high temperature. A quartz halogen lamp emits a continuous spectrum of visible light with peak emission in the IR region.

quartz objective

An objective with lenses made from fused quartz, suitable for use with ultraviolet light.

quasielastic scattering

Inelastic scattering with negligible loss of energy. Quasi-elastic scattering generates phonons.

quenching

The reversible inhibition of the emission of light by a fluorophore due to interaction with another molecule.

quick-freeze deep-etch

A technique for specimen processing comprising rapid cryofixation followed by deep etching and (rotary) replication. Quick-freeze deep-etch images reveal details of macromolecular structure and interactions in cell cytoplasm.

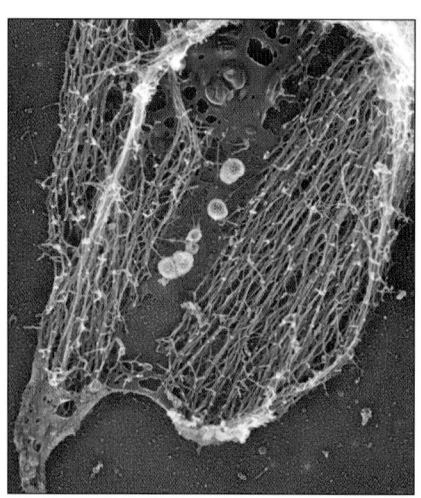

TEM image of replica of quick-freeze deep-etched cytoplasm of fibroblast. Courtesy of Julian Heath.

R

radial distribution function

A mathematical expression used in electron diffraction to describe how atoms are arranged in amorphous and crystalline materials.

radian

SI derived unit of plane angle. Symbol rad. 2π radians = 360°; 1 radian = 57.29°.

radiance

A radiometric quantity used to characterize the intensity of electromagnetic radiation. Symbol L; unit: watt per steradian per square meter.

radiant energy

A radiometric quantity used to characterize the energy of electromagnetic radiation. Symbol Q; unit: joule.

radiant flux

A radiometric quantity used to characterize the total power of a beam of electromagnetic radiation. Symbol Φ; unit: watt.

radiant intensity

A radiometric quantity used to characterize the energy of electromagnetic radiation. Symbol I; unit: watt per steradian.

radiation

Energy moving in the form of waves or particles.

radiation damage

The damage to specimens caused by an incident beam of photons or electrons. The causes and effects of radiation damage include (but are not limited to): heating; evaporation or sputtering; radiolysis; knock-on damage; photobleaching; phototoxicity; and loss of mass. In electron microscopy, radiation damage can be minimized by using very thin specimens, carbon or metal coating, minimum dose techniques, higher electron

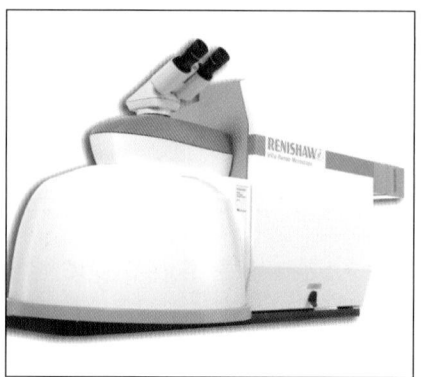

Raman microscope. Courtesy of Renishaw plc.

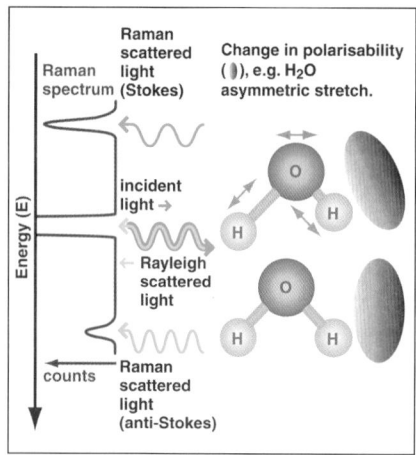

Principle of Raman scattering and Raman spectroscopy. Courtesy of Renishaw plc.

accelerating voltages or by reducing specimen temperature.

radiation pressure

The force per unit area due to the momentum of electromagnetic radiation.

radiation shielding

Materials that protect the user from harmful radiation. On electron microscopes, radiation shielding includes the use of lead glass in the viewing chamber window and lead casing of gun and magnetic lenses.

radio wave

Electromagnetic radiation with wavelengths in the range 0.3 m to several kilometers, and frequencies in the range 1 to 10^9 Hz.

radiolysis

The breakage of covalent bonds and creation of free radicals by incident radiation.

Ralph knife

A glass knife with a long blade (~25 mm) used to cut large sections from paraffin or resin blocks.

Raman-FTIR microscope

A microscope that has equipment for Raman and Fourier-transform infrared spectroscopy.

Raman microscope

A light microscope that measures and maps the Raman scattering of light by a specimen. The key features of a Raman microscope are: a laser to illuminate the specimen with monochromatic light; notch filters to block Rayleigh scattered light; and a dispersive or Fourier-transform Raman spectrometer to analyse the scattered light and produce a Raman spectrum.

Raman microscopy

The use of a Raman microscope.

Raman scattering

The inelastic (Stokes) and superelastic (anti-Stokes) scattering of ultraviolet, visible light and infrared radiation due to the polarizability of molecules, typically containing covalent bonds.

Raman spectroscopy
The characterization and quantification of the chemical components of specimens by Raman scattering.

Raman spectrum
A plot of the intensities and range of wavelengths of light emitted by a specimen illuminated with monochromatic light. In Raman spectroscopy, a Raman spectrum is compared to a reference spectral database to identify specific molecular bonds to characterize and quantify the chemical composition of the specimen.

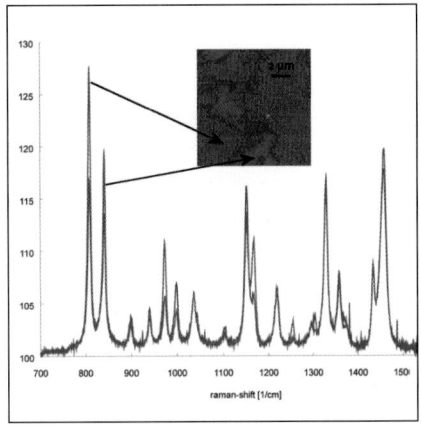

Raman map with with corresponding spectrum of a semicrystalline polypropylene film. Courtesy of WITec.

Ramsden disk see exit pupil

Ramsden eyepiece
An external diaphragm light microscope eyepiece with two planoconvex lenses.

randomly polarized light
Light waves with random planes of vibration of the electric field vector; the condition of sunlight and most microscope light sources.

rapid freezing see cryofixation

raster
A pattern of parallel lines. A raster is typically created by the motion of the beams in a video tube, scanning electron microscope or confocal laser scanning microscope.

raster image see bitmapped image

ratiometric probes
Fluorescent probes that exhibit excitation or emission spectral shifts when located in a specific chemical or physical environment, enabling the use of fluorescence ratioing.

ray
A straight line that indicates the direction of propagation of radiation and that is normal to the wavefront.

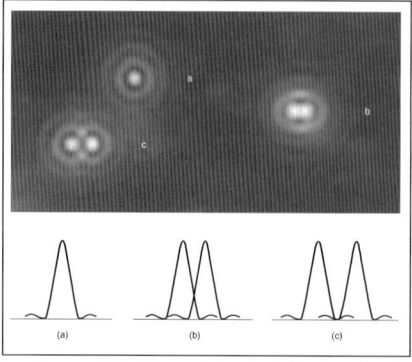

Rayleigh criterion applied to images of point sources (top) and their Airy patterns (bottom). (a) Single particle. (b) Adjacent particles just resolved. (c) Adjacent particles clearly resolved. From D. Murphy © John Wiley and Sons, Inc.

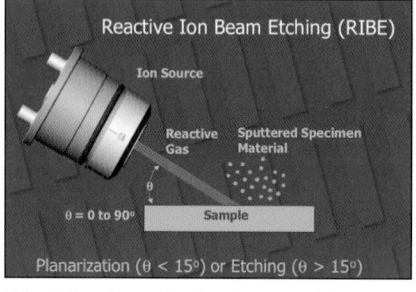

Principle of reactive ion beam etching. Courtesy of E.A. Fischione Instruments, Inc.

ray diagram

A diagram showing the paths taken by electrons or photons through a lens, microscope or imaging system.

Rayleigh criterion

A criterion for the resolving power of a diffraction-limited microscope, formulated by Lord Rayleigh. The Rayleigh criterion states that two point sources are minimally resolved when they are separated by the radius r of their Airy disks, i.e. when the center of the Airy disk of one point falls on the first minimum of the Airy pattern of the other point:
$$r = 0.61\lambda/NA$$
where λ is wavelength and NA is numerical aperture of the objective.

Rayleigh resolution see Rayleigh criterion

Rayleigh scattering

The elastic scattering of a light wave by objects sized less than one tenth its wavelength. The degree of scattering I is inversely proportional to the fourth power of wavelength λ:
$$I = 1/\lambda^4.$$
Rayleigh scattering of sunlight by air molecules is responsible for the blue sky.

reaction product

The compound that is formed by the action of an enzyme on its substrate.

reactive ion-beam etching

The etching of a surface by the combined action of a milling ion beam and a chemically reactive plasma. Typical gases used for plasma formation in RIBE are CF_4 and O_2.

reactive oxygen species

A free radical of oxygen produced by photolysis. Reactive oxygen species are a problem in fluorescence microscopy causing photobleaching.

read-out noise

Noise produced by components of the electronic processing chain of a detector, e.g. amplifier and analog-to-digital converter.

real image

An image which can be received by a viewing screen or imaging device placed in an image plane. A real image is formed on the opposite side of a lens from the object.

real-time microscopy

Any form of microscopy that allows viewing by eye or at standard video rates.

reciprocal lattice rod

The elongated form of a reciprocal lattice point arising from finite specimen thickness; if the specimen has thickness t, the relrod has length $1/t$.

reciprocal space

The three-dimensional space occupied by the diffraction pattern of a specimen, having dimensions that are reciprocal to those of the diffracting objects in the specimen.

reciprocity failure

The failure of a photographic film to obey the law of reciprocity. With reciprocity failure, film appears to become less sensitive to light when given exposures of several seconds or more. Film manufacturers publish correction factor charts for reciprocity failure correction.

rectified condenser

A condenser with strain-free non-polarizing lenses suitable for use with polarized light microscopy and differential interference contrast microscopy.

rectified optics

Optics for polarized light microscopy that have been corrected for the rotation of polarized light that occurs at oblique-incidence interfaces in optical components lying between the polarizer and analyzer.

red fluorescent protein see fluorescent proteins
red-I plate see full-wave plate

reference flat see optical flat
reference focal length see tubelength

reference viewing distance

A standard distance of 250 mm between an object and the vertex of the cornea of the eye.

reflectance

The ratio of the intensity of reflected to incident light.

reflected light microscope

A light microscope that forms an image using light reflected from the surface of the specimen. In a reflected light microscope light from an epi-illuminator is reflected by a neutral beamsplitter down through the objective, which acts as the condenser, onto the specimen. Light reflected from the specimen passes back though the objective and beamsplitter to form the intermediate image.

reflected light microscopy

The use of a reflected light microscope.

reflected light mode see episcopic mode
reflected-light interference microscope see interference microscope
reflecting sphere see Ewald sphere

reflection

1. The return of incident light at an interface.
2. A region of high intensity in a diffraction pattern.

reflection contrast microscopy

The use of a reflected light microscope with polarized optics to enhance the contrast of images of weakly reflecting objects such as stained and cytochemically labeled specimens.

reflection electron energy-loss spectroscopy

Electron energy-loss spectroscopy of electrons reflected from the surface of a specimen.

reflection high-energy electron diffraction

The production of a diffraction pattern by reflection of high-energy electrons from the surface of a specimen.

reflection interference contrast see
reflection contrast microscopy
reflection mode see **episcopic mode**
reflection, law of see **law of reflection**

refraction

The bending of light at an interface between two
materials of different refractive indices when the
angle of incidence is greater than 0°. Refraction is
caused by the change in the velocity of light in the
two media. The direction changes in accordance
with the law of refraction.

refraction, law of see **law of refraction**

refractive index

A measure of the speed of light in a transparent
medium. Symbol n. Refractive index is defined as
the ratio of the speed of light c in a vacuum to its
speed v in a given material:

$$n = c/v.$$

Thus the refractive index of a vacuum is 1. The
refractive indices of common media are: air =
1.0003; water = 1.33; quartz glass = 1.54.

refractive-index matching

The placement of a transparent object in a medium
with similar or identical refractive index.
Refractive index matching may be used to measure
refractive index, or to reduce the degradation of
images, e.g. to reduce phase halos in phase
contrast light microscopy.

refractive-index mismatch

Changes of refractive index in the light path
between a plane of focus in a specimen and the
objective lens of a light microscope. Since light
microscope objectives are corrected for light paths
through media of uniform refractive index,
refractive index mismatch arising from the
immersion medium, coverslip, and material
surrounding and within the specimen can cause
degradation of images.

refractometry

The measurement of refractive index.

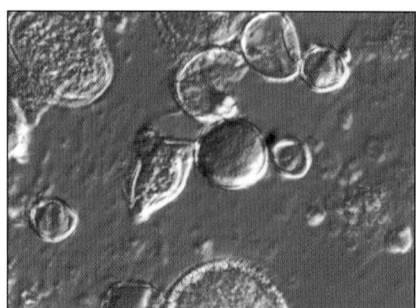

Relief contrast microscopy. Courtesy of Olympus.

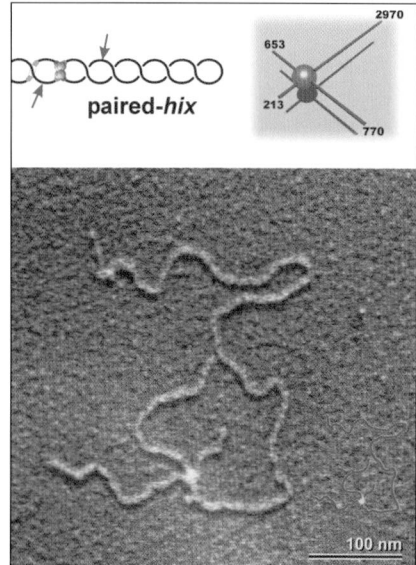

TEM image of replica of DNA. *Microscopy and Analysis* 2000.

relative retardation

The optical path difference between the E and O rays in an optically anisotropic material. Symbol Γ. Relative retardation is given by:

$$\Gamma = (n_e - n_o)t$$

where n_e and n_o are refractive indices for the E and O rays respectively and t is thickness.

relativistic accelerating voltage see accelerating voltage
relativistic wavelength of electron see electron

relief contrast microscopy

A proprietary design of Hoffmann modulation contrast light microscopy.

relrod see reciprocal lattice rod

remote electron beam-induced current

A variant of electron beam-induced current imaging suitable for studies on resistive materials, such as electroceramics. In REBIC imaging, two widely spaced (hence remote) contacts are placed on the sample; one is grounded and the other is coupled to an amplifier and produces the EBIC image.

remote microscopy see telemicroscopy

replica

A thin coating that reproduces the surface topography of a specimen. The replica is typically formed by carbon, metal or polymers and may be detached from the specimen before examination. A replica is commonly used in scanning and transmission electron microscopy for the study of bulk, opaque or weakly scattering specimens.

replication

The production of a replica.

resin

A material for specimen embedding. On polymerization, resins form a hard matrix around and inside specimens facilitating specimen processing by microtomy or grinding and polishing. Commonly used resins include the epoxy resins, methacrylates and acrylic resins.

resin removal

The removal of resin from sections of an embedded specimen in order to reveal structure or binding sites for cytochemical probes. Polymerized resins can be dissolved in sodium ethoxide, sodium methoxide or organic solvents.

resolution

The amount of detail in an image.

resolution standards see calibration standards

resolving power

The ability of a lens or optical system to discriminate detail in a specimen. Resolving power is a function of the wavelength of the illuminating radiation and the angular separation of points in the specimen. The term resolving power is widely used synonymously with resolution, which is the discrimination of detail by an image.

resolving power, of light microscope

The resolving power of a light microscope objective is a function of the wavelength λ of the illumination and the numerical aperture NA of the objective lens. For a diffraction-limited objective, the minimum resolvable distance d is given by:
$$d = 0.612\lambda/NA.$$

resonance

The oscillation of a material, molecule, atom or system at its natural frequency of vibration. At resonance, large amplitude can result from minimal energy input.

resonance ionization imaging detector

A detector that uses laser resonance ionization of an atomic vapor, e.g. mercury, to amplify input signals. Photons entering the detector ionize the atoms of a laser-excited metal vapor; the metal ions are then accelerated through a microchannel plate onto a phosphor screen which is imaged with a CCD.

resonant frequency, of cantilever

The frequency of vibration of a cantilever in a scanning probe microscope. Symbol: f_o; unit: Hz. For a beam cantilever:

$$f_o = 0.162(E/\rho)^{1/2} \times t/l^2$$

where E is Young's modulus, ρ is density, t is thickness, and l is length. For standard cantilevers, f_o is in the range 1-400 kHz.

rest mass of electron see electron

resultant wave

The wave formed by the interference of two or more waves.

retardation

The reduction in the speed of a wave on entering a medium of higher refractive index.

retardation plate

A transparent plate made from optically anisotropic material that produces a fixed optical path difference or retardation between transmitted E and O rays. The optic axis of the material is parallel to the upper and lower surfaces of the plate.

retarder see retardation plate

reticle

A glass disk with inscribed scale or orientation markings that is placed in an eyepiece, in the intermediate image plane. When used in combination with a stage micrometer, a reticle calibrates magnifications and measures objects in the field of view.

reticule see reticle

retina

The light-sensitive inner lining of the posterior chamber of the eye. The human retina contains approximately 120 million rod cells and 6 million cone cells that detect intensity and color respectively. The retina is the image plane of the eye.

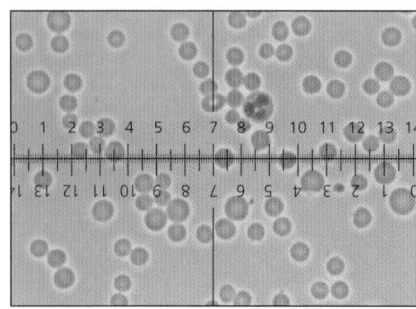

Scale on a reticle. Courtesy of Carl Zeiss Ltd.

reverse-biased p-n diode

A p-n diode having an applied voltage that prevents the flow of current.

reverted image see erect image

RGB

1. Abbreviation for red, green and blue, three additive colors that make white.
2. A standard image format used in electronic imaging and detectors.

Rheinberg illumination

A type of darkfield light microscopy in which the specimen appears in one color on a differently colored background. For Rheinberg illumination, a filter comprising a colored disk with a differently colored annulus is placed in the front focal plane of the condenser; the annulus provides the oblique illumination that is diffracted by the specimen and the complementary area of the filter provides the direct background illumination.

ribbon see section ribbon
Riecke diffraction see nanodiffraction

right-circularly polarized light

Circularly polarized light in which the electric field vector rotates clockwise (as seen by an observer looking towards the source).

ring light

A source of illumination in the shape of a ring, containing a fluorescent lamp or a circular array of light-emitting diodes.

Robinson detector

A backscattered electron detector commonly used in scanning electron microscopes. The Robinson detector comprises an annular scintillator coupled via a light pipe to a photomultiplier tube.

rod cell

A photoreceptor cell in the retina involved in non-color and low-light vision. Rods are 100 times more sensitive to light than cones.

Ring lights. Top: Fluorescent ring lights. Bottom: Light-emitting diode ring lights. Courtesy of StockerYale Inc.

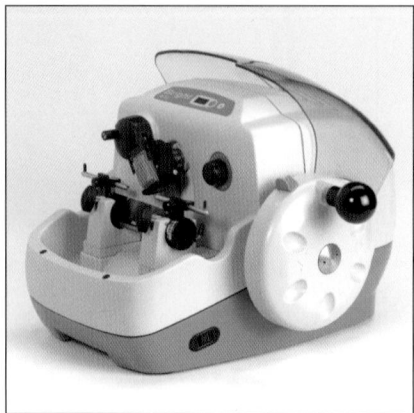

Rotary microtome. Courtesy of Bright Instruments.

roentgenoluminescence
Luminescence produced by X-rays.

Ronchigram
An image generated by the interference of overlapping diffraction spots or disks. Ronchigrams are used in the testing and aberration correction of optical and electron imaging systems. In electron microscopy a Ronchigram interference pattern is formed in coherent CBED mode when the illumination angle is very large. This aberration figure can be analysed to give the aberration constants of a STEM objective lens, and so has become the preferred method of aligning a STEM for highest image resolution

rotary coating see rotary replication

rotary microtome
A standard type of microtome used in histology in which the cutting cycle is controlled manually or semi-automatically by a wheel mounted on the side of the microtome.

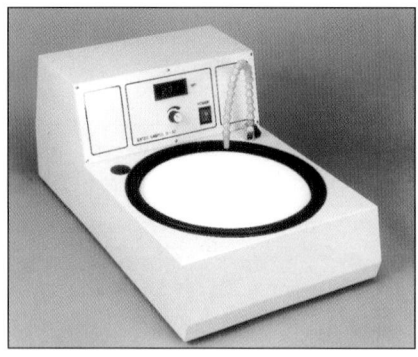

Rotary polisher. Courtesy of Agar Scientific.

rotary polisher
A device used for wet grinding and polishing of materials specimens. The specimen is held in a grinding jig or tripod polisher and placed on a variable speed rotating platen in the presence of abrasive grits.

rotary pump
A mechanical pump that performs the preliminary evacuation of electron microscopes and other vacuum devices. A rotary pump has a pump chamber containing an eccentric rotor with spring-loaded vane seals creating two compartments. In one cycle, air is successively drawn in to each compartment from the microscope column, compressed and expelled through the outlet valve, creating a vacuum that draws in more air for the next cycle. Oil is used to lubricate the chamber lining and must be prevented from contaminating the microscope or environment with oil traps. Rotary pumps produce a vacuum of around 10^{-1} Pa.

Rotary pump. Courtesy of Agar Scientific.

rotary replication
The production of a replica while rotating the specimen. Rotary replication improves access of the replica material to the specimen.

rotary shadowing
The shadowing of a rotating specimen.

rotary vane pump see rotary pump

rotating stage
A light microscope stage that is rotatable by 360°, and used for polarization microscopy or in any application where the orientation of the specimen relative to either the illumination or the imaging device is important.

rotation-free imaging see image rotation in EM
roughing pump see rotary pump
round lens see round magnetic lens

Rowland circle
The circle of fixed diameter whose circumference passes through the source, diffracting crystal and detector in an optical or X-ray spectrometer. aka focusing circle. The Rowland circle is based on the geometrical principle that an arc (the crystal) subtends the same angle at any two points (the source and detector) on its circle. Since the source is invariably fixed, maintaining the crystal and detector on the Rowland circle as they are moved ensures that any beam diffracted by the crystal is always in focus at the detector.

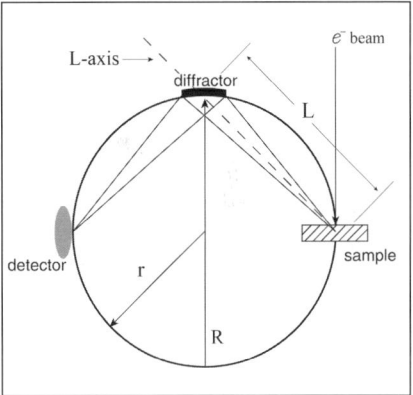

Geometry of Rowland circle in a spectrometer. Courtesy of Thermo Electron Corporation.

Rutherford backscattering see Rutherford scattering

Rutherford scattering
Elastic scattering at large angles, up to 180°. Rutherford scattering occurs as a result of collisions with atomic nuclei.

S

safety light

A low intensity red or orange light used when handling photographic film and paper in a darkroom. Most slow-speed film plates used for transmission electron microscopy can be handled under a safety light; roll film should be handled in complete darkness.

sagittal focus

A focal plane at a sagittal plane of a specimen.

sagittal plane, sagittal section

A plane or section that divides an object vertically and longitudinally into left and right halves (mid-sagittal) or two unequal portions.

sapphire knife

A knife made from sapphire used to cut thin sections in an ultramicrotome.

saturation

The degree to which a color departs from white light.

scale bar

A line placed in an image to indicate a specified distance in the object.

scan coils

A set of double-deflection coils, located above the condenser-objective lens, that scan the electron beam across the specimen in SEMs, TEM-STEMs, and STEMs.

scan head

A component of a confocal or scanning microscope that generates the scanning beam.

scan lens

In a confocal microscope, a lens that projects an image of the scanning beam into the back focal plane of the objective.

scanner

An instrument, device or mechanism that moves a probe over an object. In microscopy and digital imaging the term scanner is commonly used to describe those devices that: 1. Move the probe in a scanning probe microscope. 2. Move a light beam in a confocal or laser microscope. 3. Move the beam in an electron microscope. 4. Scan a document or photograph.

scanning acoustic microscopy

The use of an acoustic microscope with a scanning beam of ultrasound waves.

scanning Auger microscopy

The use of a scanning electron beam in Auger electron spectroscopy.

scanning capacitance microscopy

The use of a scanning probe microscope to measure and map the capacitance between probe and specimen. In SCM a conductive tip is scanned in contact mode over the surface of the specimen; an ac voltage bias applied between tip and specimen generates capacitance variations which are detected by a capacitance sensor. A typical application is the mapping of dielectric thickness and carrier profiles in semiconductors.

scanning coils see scan coils

scanning electrochemical microscopy

The use of a scanning probe microscope to map and measure changes in the faradaic current between the probe and a specimen. In SECM an ultramicroelectrode or ion-selective electrode is immersed in an electrolyte and measures changes in faradaic currents flowing at the probe tip, due to electrolysis of species in the electrolyte, as a function of distance from the specimen. When the tip-specimen gap is very small, a non-conducting specimen surface reduces tip current due to physical restriction of diffusion of electrolyte species, whereas a conducting surface allows currents to continue to flow. The current is used for image formation. SECM may be used for surface modification.

Scanning electron microscope. Courtesy of FEI Company.

scanning electrochemical potential microscopy

The use of a scanning probe microscope to map and measure the electrochemical potential of a specimen. In scanning electrochemical potential microscopy a conducting tip, immersed in an electrolyte or polar fluid, measures the potential difference between the probe and the specimen. The potential is used for image formation. Typical applications are the study of electroplating and corrosion.

scanning electron microscope

An electron microscope that scans a focused beam of electrons across a specimen. The key components of an SEM are: a thermionic or field emission electron gun; beam scanning coils; one or more condenser/objective lenses; a large specimen chamber with a stage for bulk specimens; detectors for imaging and spectroscopy; and a cathode ray tube for image display. A scanning electron microscope has no post-specimen lenses. The primary method of image formation is by collection of secondary and backscattered electrons. The scan controls of the electron beam and CRT are synchronized; as the beam scans the specimen the output from the secondary or backscattered detector is coupled to the CRT voltage producing contrast on the CRT display monitor. Compositional data are obtained by analysis of the electrons, light and X-rays emitted by the specimen.

scanning electron microscopy

The use of a scanning electron microscope.

scanning force microscope see atomic force microscope

scanning force microscopy

A collective term for a family of scanning probe microscope techniques that measure and map the force between a probe and a specimen.

scanning Hall probe microscopy

The use of a scanning probe microscope equipped with a miniaturized Hall probe to measure and map magnetic fields at the surface of a specimen.

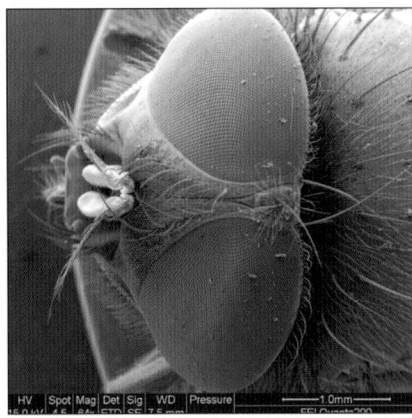

Scanning electron micrograph of fly head. Courtesy of FEI Company

scanning head see confocal head

scanning impedance microscopy

The use of a scanning probe microscope to map and measure the ac and dc transport properties such as impedance, capacitance and resistance of a specimen.

scanning ion-conductance microscopy

The use of a scanning probe microscope to measure and map the ion conductance between an electrode in a micropipette and a specimen immersed in an electrolyte. As the micropipette approaches the specimen surface the cross-section of the ion path reduces and conductance decreases; the tip-sample separation is kept constant by controlling the ion current. SICM may be used for profilometry of living cells and porous materials.

scanning Joule-expansion microscopy

A form of scanning thermal expansion microscopy in which Joule heating is applied to an electrically conducting specimen.

scanning Kelvin probe microscopy see scanning surface potential microscopy

scanning laser microscope

A microscope that uses a scanning laser beam as the source of illumination.

scanning local-acceleration microscopy

A mode of force modulation microscopy in which the specimen position is modulated at a frequency above the highest tip-sample system resonance.

scanning microscopy

Any form of microscopy in which a probe is scanned across the specimen (or *vice versa*).

scanning near-field acoustic microscopy

The use of a scanning probe microscope to measure and map the interactions of acoustic (sound) waves with the specimen in the near-field.

scanning near-field optical microscope see near-field scanning optical microscope

SEM image of a 170-nm wide bismuth Hall probe on a cantilever tip. Courtesy of Dorothee Petit, Univ. Durham, UK

scanning photoacoustic microscope

The use of a scanning laser beam to induce sound waves in a specimen that are detected by a piezoelectric crystal.

scanning probe microscope

A microscope that scans a sharp probe very close to the surface of the specimen to obtain topographical, chemical or physical data at very high resolution. The key components of a scanning probe microscope are: a sharp probe or tip, typically at the end of a cantilever; a scanning system, usually a piezoelectric scanner, to move the probe relative to the specimen, or *vice versa*, at nanometer resolution; a probe motion sensor to determine the position of the probe (usually either an optical system, using a laser beam reflected from the cantilever, or a piezoresistive detector); an electronic controller system that regulates the probe-specimen interactions and receives the image-forming signals; a computer to drive the system and process the images; a light microscope to examine the specimen and probe; and a robust frame to reduce vibration.

scanning probe microscopy

A collective term for a family of microscopical techniques that have in common the use of a probe that is scanned in very close proximity to the surface of a specimen yielding images and data on the chemical, physical and topographical properties of a specimen at very high resolution.

scanning resistance microscopy see scanning spreading-resistance microscopy

scanning spreading-resistance microscopy

The use of a scanning probe microscope to measure and map the electrical conductivity or resistivity of a specimen. In SSRM a conductive probe is scanned in contact mode over the specimen surface, measuring electrical resistance between the tip and a back contact attached to the specimen; a hard probe tip (e.g. doped diamond) is required for the high forces used to establish good electrical contact with the specimen.

scanning stage

A programmable motor-driven light microscope stage that scans the specimen(s) under the beam.

scanning surface potential microscopy

The use of a scanning probe microscope to map and measure the electrostatic potential at the surface of a specimen. In SSPM a conductive biased tip is used to detect the electrostatic potential. Typically a two-pass method is used: first the topography of the specimen is imaged; the tip is then moved away from the specimen and a second scan is used to detect electrostatic charges at the specimen surface. These charges cause the cantilever to experience a force if the tip has a different potential than the region of the specimen below it; the force is nullified by varying the voltage of the tip so that it is at the same potential. The applied voltage is used to generate the image.

scanning thermal expansion microscopy

The use of a scanning probe microscope to measure and map the expansion of a specimen by heating, typically using a resistive thermal probe.

scanning thermal microscopy

The use of a scanning probe microscope to map and measure the thermal properties of a specimen. In scanning thermal microscopy a thermal probe acts as the detector, and sometimes the source, of temperature changes in the specimen.

scanning transmission electron microscope

An electron microscope that scans a highly focused electron beam across a thin specimen generating images and spectroscopic data from the electrons transmitted through the specimen. An STEM requires no post-specimen lenses so images do not suffer from the aberrations found in a TEM. Images are formed from scattered electrons collected by annular detectors beneath the specimen. There are two configurations of STEM. Dedicated STEMs instruments are usually configured as an inverted SEM with a field-emission electron gun at the base of the column, for stability, and imaging and analytical detectors at the top. A dedicated STEM has no post

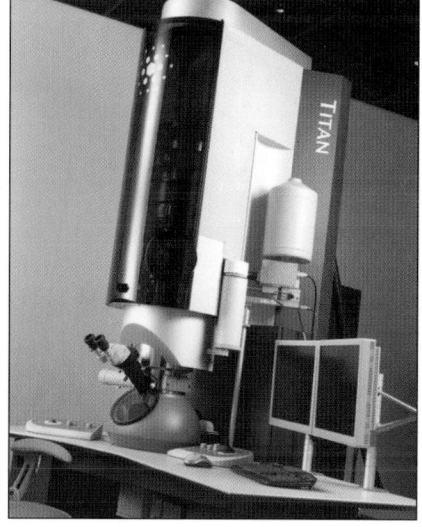

Scanning transmission electron microscope (TEM-STEM). Courtesy of FEI Company.

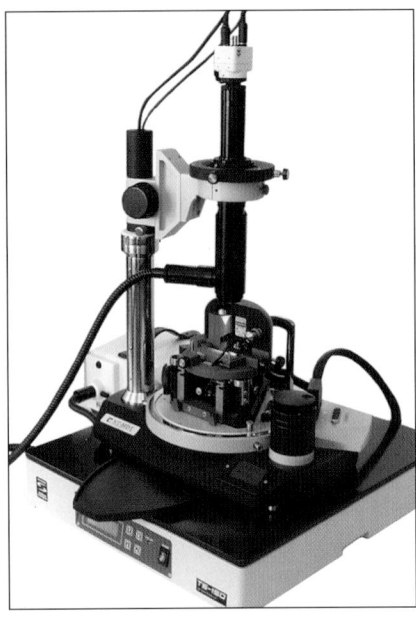

Scanning tunneling microscope. Courtesy of NT-MDT.

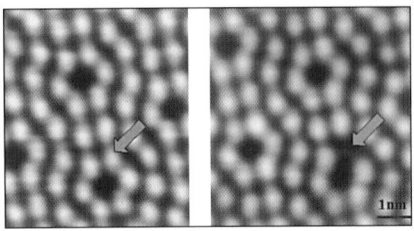

Scanning tunneling microscope imaging and manipulation of silicon atoms. *Microscopy and Analysis* 2001.

specimen lenses. TEM-STEMs are transmission electron microscopes equipped with beam scanning coils and annular detectors.

scanning transmission electron microscopy

The use of a scanning transmission electron microscope.

scanning transmission X-ray microscope/y

An X-ray microscope in which the specimen is scanned raster fashion through the beam, building an image pixel by pixel.

scanning tunneling microscope

A type of scanning probe microscope that measures and maps the electron tunneling current between a probe and the specimen. In an STM, the probe is an ultrasharp conductor that is placed 0.1 to 5 nm from the surface of a conducting or semiconducting specimen; an applied voltage between the probe and specimen allows electrons to cross the energy barrier between the probe tip and the specimen resulting in a tunneling current from tip to specimen (or vice-versa, according to the sign of the applied voltage). The tunneling current is exponentially proportional to the tip-to-sample distance; a 0.1 nm decrease in tip-specimen distance increases the current by an order of magnitude, which gives the STM its high vertical sub-nanometer resolution. Images or surface maps are formed either by using constant current or constant height modes: in the former mode, the signals required to alter the probe position provide the image dataset; in the latter, the changes in tunneling current are used to form the image.

scanning tunneling microscopy

The use of a scanning tunneling microscope.

scanning-slit confocal microscope

A type of scanning confocal microscope that uses slit apertures instead of pinholes in the laser and detector optical pathways.

scattering

The change in the direction of propagation of electromagnetic radiation or charged particles as a

scattering cross-section

result of their interaction with matter.

scattering cross-section
The cross-section of an atom that leads to an elastic or inelastic scattering event. Symbol σ; unit: barn.

Scherzer defocus
The optimum amount of defocus (underfocus) for the maximum information transfer in high-resolution electron microscopy. Scherzer defocus Δf is given by the expression:
$$\Delta f = -1.2(Cs\lambda)^{1/2}$$
where Cs is the spherical aberration coefficient and λ is wavelength.

Scherzer focus see Scherzer defocus

Scherzer's theorem
Scherzer's theorem states that chromatic and spherical aberrations can never be eliminated from round magnetic lenses by skillful design.

Schlieren contrast microscopy
A form of modulation contrast microscopy commonly used to examine differences in density or refractive index. In Schlieren contrast microscopy the object is illuminated by a slit aperture; a conjugate opaque modulator in the back focal plane of the objective blocks some of the zero-order light and all of one sideband of the diffraction pattern. Image contrast derives from interference in the image plane of the remaining zero-order and higher order diffracted light in the non-blocked sideband.

Schottky field emitter
A thermal field emitter used in electron guns, that typically has a tip coated with ZrO_2 and is heated to ~1500°C.

Schottky field-emission
Field emission that is enhanced by heating.

scintillator
A luminescent compound used in a variety of detectors. Scintillators may be made of inorganic phosphors or polymers such as vinyltoluene.

screen lift lever

A lever at the side of the viewing chamber of a TEM that moves the focusing screen to a 45° angle for viewing or to a vertical position for photography.

scroll pump

A type of dry rotary pump that uses two spiral-shaped plates, one rotating around the other, that repeatedly compress and decompress the gas or vapour to be pumped, expelling them through the outlet at the center of the fixed scroll.

SECAM video

The systéme electronique couleur avec mémoire video system, a 625-line, 25 frames per second color video broadcast standard used in Europe.

second

SI base unit of time. Symbol s. One second is the duration of 9,192,631,770 periods of the radiation produced by the transition between two hyperfine levels of the ground state of the cesium-133 atom.

secondary electron

An electron released by a specimen by an incident electron beam. Secondary electrons are classified by their source and energy: slow secondary electrons are ejected from conduction or valence bands and have energies below 50 eV; fast secondary electrons are ejected from inner shells and have energies of up to 50% of the incident beam.

secondary electron detector

A detector of secondary electrons. A standard design is the Everhart Thornley detector used in many scanning electron microscopes. Environmental SEMs use a gaseous secondary-electron detector; variable-pressure SEMs use a variable-pressure secondary electron detector.

secondary electron imaging

The use of secondary electrons to form an image, typically in a scanning electron microscope.

secondary emission

Emission of a photon or particle following

excitation from an emitted photon or particle.

secondary extinction

The reduction in intensity of a diffraction pattern due to weaker penetration of the primary beam into deeper layers of the specimen.

secondary fluorescence

Fluorescence produced by excitation from radiation generated within a specimen or detector. Secondary fluorescence may occur in fluorescence microscopy and X-ray microanalysis.

secondary ion

An ion emitted by a material as a result of ion beam bombardment. Secondary ions are detected in secondary ion mass spectrometry.

secondary-ion mass spectroscopic imaging

The use of secondary ion mass spectrometry to map and measure elements in a specimen. In SIMS imaging, the specimen is scanned with a micrometer-diameter focused beam of charged particles (typically Ga^+) at 1-20 keV. Secondary ions sputtered from a 1-nm deep surface layer are analyzed with a mass spectrometer to produce a map.

secondary X-ray

An X-ray emitted as a result of excitation with a primary X-ray.

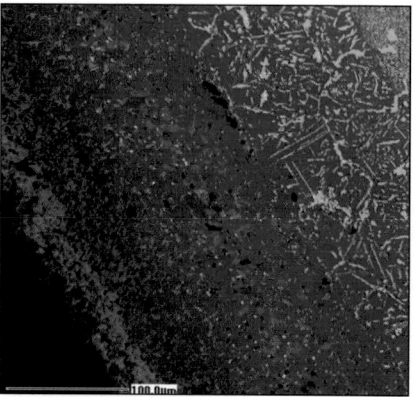

Secondary ion mass spectroscopic image of of nickel superalloy: Al red, Ti green, Cr blue. *Microscopy and Analysis* 2000.

second harmonic generation

A non-linear optical phenomenon in which two identical incident photons combine in a medium to form a single photon with a frequency double that of the interacting photons.

second harmonic microscopy

A mode of light microscopy that uses second harmonic generation photons to image specimens. Second harmonic microscopy can detect differences in specimen physical properties at interfaces, such as immersion oil and glass, or cell and water, which cannot be seen with normal optical microscopy, and provides optical sectioning of thick specimens.

second-order Laue zone see **Laue zones**

section
In microtomy, a slice of finite thickness through a specimen.

sectioning angle
The angle between the block face and the rear surface of a microtome knife. The sectioning angle is the sum of the clearance angle and the included angle.

section ribbon
A ribbon formed by a consecutive set of serial sections.

section thickness estimation
The thickness of sections and foils can be estimated by using: 1. interference colors; 2. stereopairs; 3. thickness mapping; 4. embedding and cross-sectioning.

section thickness measurement
The thickness of microtome sections may be estimated from: 1. The microtome's section-thickness control. 2. Interference colors of thin sections. 3. Minimal folds. 4. Stereoscopy of internal or applied markers on either side of the section. 5. Cross-sections of re-embedded sections. 6. Mass thickness.

Seidel aberrations see **monochromatic aberrations**

selected-area diffraction aperture
An aperture (diaphragm), placed in the first intermediate image plane of a transmission electron microscope, used to select the area of a specimen used to form a diffraction pattern and in low magnification imaging.

selected-area diffraction see **selected area electron diffraction**

selected-area electron diffraction
Electron diffraction using a parallel beam of electrons and selecting a small area ($\sim 1\mu m^2$) of the

specimen for study using a diffraction aperture. Selected area electron diffraction is the basic mode for electron diffraction in a transmission electron microscope.

self-biasing gun

An electron gun in which the filament current regulates the potential difference (bias) between filament and Wehnelt ensuring stable emission and prolonging the life of the filament. In a self-biasing gun, the negative high voltage supply is connected to the Wehnelt and to the filament through a variable bias resistor. This arrangement creates a small potential difference (~200 V) between Wehnelt and filament, the gun bias voltage. As the filament heating current increases, current flows from the filament through the bias resistor to the Wehnelt causing the bias voltage to increase, so reducing the emission from the filament.

semiapochromat

A light microscope objective intermediate in correction between an achromat and an apochromat, having lens(es) made from fluorospar. Semiapochromats have a broad wavelength transmittance, from ~130 nm to ~10 μm.

semiapochromatic objective see semiapochromat

semiconductor

A crystalline solid (e.g. silicon or germanium) with an electrical conductivity intermediate between that of an insulator and a metal conductor. The conductivity is temperature dependent: at 0K semiconductors behave as insulators; at room temperature thermal excitation allows electrons to move from the valence band into the conduction band leaving holes (positive charges) in the valence band. Under an applied electric field conduction occurs as a result of net movement of electrons in the conduction band and holes in the valence band. An intrinsic semiconductor is one in which the concentration of electrons and holes (the charge carriers) is a property of the semiconductor material. For most electronic applications, semiconductors are doped with other elements. An

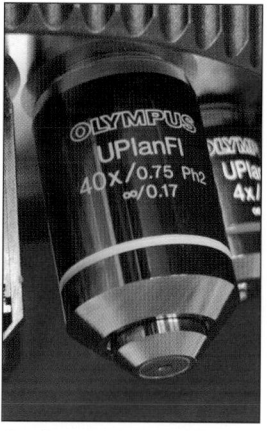

Semi-apochromat objective. Courtesy of Olympus.

287

extrinsic semiconductor is one that has additional charge carriers: a n-type conductor has additional negative charges, typically by doping with boron; a p-type has extra positive charges, typically from arsenic or phosphorus. A p-n junction can be created by doping adjacent regions of a crystal with p- and n-type dopants, creating a p-n diode. When a positive voltage bias is placed on the p-type side holes and electrons are moved towards the junction and current can flow; if the bias is reversed, the junction becomes depleted of electrons and holes and no current flows.

semiconductor detector

A detector that uses a semiconducting material to detect photons and charged particles.

SEM Raman system

A Raman spectroscopy system attached to a scanning electron microscope allowing the acquisition of complementary Raman spectra and scanning electron microscopy images and data.

SEM-STEM see scanning transmission electron microscope

semithin section

A section 200-500 nm thick, typically cut for light microscope evaluation of a block prior to the production of ultrathin sections.

sensitive tint plate see full-wave plate

serial collection mode

In WDX and EELS, the acquisition of a complete spectrum by serial analysis of each wavelength or energy.

serial electron energy-loss spectrometer

An electron spectrometer designed for serial acquisition of an electron energy-loss spectrum. The spectrometer is equipped with a slit aperture in the dispersion plane allowing electrons of a specific energy to pass through to a photo-multiplier. By ramping the current in a set of deflection coils in front of the aperture, or changing the accelerating voltage, an entire energy-loss spectrum may be acquired serially.

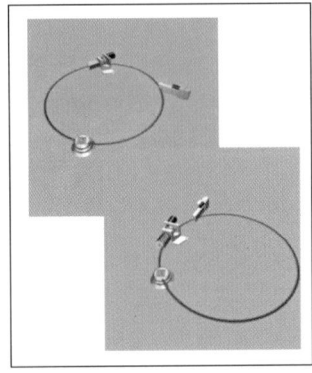

Serial collection mode in WDXS. The images show the orientation of a wavelength-dispersive X-ray spectrometer for two different wavelengths. Courtesy of Oxford Instruments.

serial electron energy-loss spectroscopy

The original, now rarely used form of electron energy-loss spectroscopy in which each channel of an EELS spectrum is acquired consecutively by selecting a specific energy-loss bandwidth with the detector or electron microscope controls.

serial register

A single row of pixels at the edge of a CCD that receives the charges from the parallel register and passes them serially to the output node.

serial sections

A set of consecutive optical or microtome sections of a specimen. Images of serial sections are typically used in 3D reconstruction.

sextupole

An electron optical device with six electromagnetic coils (poles) arranged in a circle around the optic axis, typically used for aberration correction in electron microscopes and spectrometers.

shading

The variation in amplitude of output signal when an imaging device is illuminated uniformly.

shadow image

An image formed by a (point) projection microscope. A point-projection image can be formed a scanning transmission electron microscope with the illumination aperture removed and the probe focused before or after the sample.

shadowing

The formation of a replica by application of replica material at an oblique angle to the surface of the specimen. Shadowing reveals subtle topographical features as the coating material is trapped by raised features and excluded from sunken ones giving differential deposition of material.

shape analysis

A procedure used in image analysis to quantify the parameters that describe the shape of an object, such as length, area, orientation, etc.

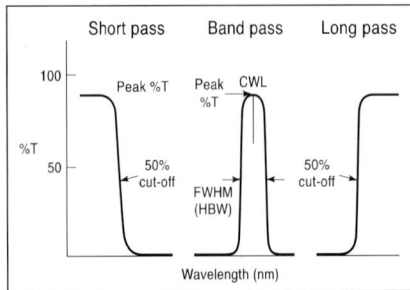

Principle of short-pass filter. From D. Murphy © John Wiley and Sons, Inc.

Motor-driven shutter for light microscopy. Courtesy of Prior Scientific Instruments.

Shuttle stage. Courtesy of Prior Scientific Instruments.

shear distance

The lateral distance between the O and E rays leaving a beam splitting prism.

shear force microscopy see transverse dynamic force microscopy

shear force mode

A mode in scanning probe microscopy in which the the tip oscillates parallel to the specimen surface.

short-arc lamp

A high-pressure gas-discharge lamp with an arc length that is small compared to the size of the electrodes.

short-pass filter

A filter that transmits wavelengths shorter than a certain specification and blocks longer wavelengths.

shot noise

The variation in output from a detector caused by random changes in the number of incident photons in unit time. Shot noise is an inherent feature of all detectors since it derives from the quantum nature of light. Shot noise is proportional to the square root of signal value.

shutter

An opaque object that blocks, or a mechanism that deflects, the passage of a beam through an optical system or into an imaging device or spectrometer.

shuttle stage

A motor-driven light microscope stage used for the loading and inspection of specimens.

Si(Li) detector see lithium-drifted silicon detector

side-entry camera

A film or CCD camera designed to be placed at the side of the column of a transmission electron microscope. A side-entry camera has a pneumatic mechanism to introduce the detector or film into the path of the electron beam.

side-entry holder

A specimen holder that is inserted between the polepieces of the objective lens of a transmission electron microscope.

sidewall imaging

Use of long, narrow probes to image the sides and base of trenches with a scanning probe microscope.

signal amplification

1. The amplification of electronic signals.
2. In cytochemistry, the use of intermediate labels (bridges) between the primary probe and the detected probe.

signal enhancement of gold probes see autometallography

signal-to-noise ratio

The ratio of the signals leaving the specimen to the noise produced by the signal detector.

silicon detector

A semiconductor detector made of a single crystal of silicon, used in charge-coupled devices and energy-dispersive spectrometers.

silicon drift detector

A type of Si(Li) detector with concentric electrodes that steer the collected charges towards a centrally placed field-effect transistor. A silicon drift detector can operate efficiently without cooling to liquid nitrogen temperatures.

Silicon drift detectors. Courtesy of Thermo Electron Corporation and Ketek.

silicon film

A support film made of silicon monoxide or silicon nitride.

silicon intensifier-target camera

A two-stage video-tube camera with a silicon diode detector between the front target and the electron beam. Electrons generated at the target are accelerated towards the silicon diode generating additional electrons. An intensified silicon intensifier-target camera (ISIT) is a SIT with an image intensifier in front of the target.

silver autometallography

A mode of autometallography that enhances the contrast of metal probes by the addition of silver atoms.

silver enhancement see silver autometallography

simple eyepiece

An eyepiece that contains only one lens, the eyelens.

simple lens see thin lens

simple microscope

A light microscope that has only one image-forming lens, the objective lens.

sine condition

The sine condition is expressed as:
$$n_o \sin\theta_o / n_i \sin\theta_i = M$$
where n_o and n_i are the refractive indices in object and image space respectively, θ_o and θ_i are the angles in these spaces, and M is magnification.

single-prism magnetic spectrometer

An electron spectrometer containing a single magnetic prism.

single-layer antireflection coating

An antireflection coating that has a thickness of one quarter wavelength, or odd multiples thereof, of the reflected light.

single-polepiece lens

A magnetic lens with single operational polepiece, such as the snorkel lens. These lenses do have a second polepiece but it is far from the optical axis.

single scattering

The scattering of an electron or photon as a result of interaction with one atom.

single-sideband edge enhancement

A type of modulation contrast microscopy that

generates phase contrast images by blocking light
in one sideband of the diffraction pattern.

single-tilt holder
A specimen holder that allows tilting of the
specimen about one axis (usually the longitudinal
axis of the holder).

sintered-glass diffuser
A diffuser made of sintered glass particles.

slam freezing see metal mirror cryofixation

slide
A thin piece of material (usually rectangular and
made of glass) used to support specimens or
sections for light microscopy and histology.

sliding microtome
A microtome equipped with a gliding sledge to
move the specimen over the knife edge.

sliding stage
A light microscope stage comprising two plates;
the upper plate carrying the specimen can be
moved laterally by hand over the fixed lower plate.

slit scanning
A mode in confocal microscopy in which an
illuminated slit instead of a pinhole aperture is
scanned across one axis of the specimen, allowing
real-time imaging.

slot grid
An electron microscope support grid with a central
slot, used for unobstructed viewing of large
sections or serial sections.

slow axis
The axis of an optically anisotropic material along
which light travels at slowest velocity.

slow-scan CCD
A type of CCD that acquires images at rates lower
than standard video rates, typically to improve
sensitivity and reduce read noise.

slow secondary electron see **secondary electron**

small-angle elastic scattering

The elastic scattering of electrons or photons through small angles, i.e. less than 5°, due to interaction with the electron shells of an atom. Small-angle elastic scattering is the major type of scattering used for image formation in darkfield electron microscopy and in electron diffraction.

smear

The high intensity streaks extending from bright objects in a video image. Vertical smear occurs in an interline-transfer CCD image when unwanted charges form in the storage areas as a result of scattering of light from illuminated pixels or leakage from the parallel register. Comet-tail smear occurs when there is persistence of signal in the target of a video-tube camera or on a phosphor screen.

Smear in interline-transfer CCD image. Courtesy of PCO AG.

Snell's law of refraction see **law of refraction**

snorkel lens

A round magnetic lens with a conical polepiece that protrudes below the body of the lens, producing a snout or 'snorkel'.

soft X-rays

X-rays with wavelengths of 1-10 nm and energies of 100-1000 eV, typically used for the imaging of hydrated biological specimens.

solid angle see collection angle

solid-state detector

A colloquial term for a detector that uses semiconductor technology, such as a charge-coupled device.

sonoluminescence

Luminescence produced by sound waves.

sound wave

A longitudinal or transverse pressure wave that can be detected by the human ear. Sound waves lie in the frequency range 20-20,000 Hz, and travel (at 20°C) at a speed of 344 m s^{-1} in air, 1482 m s^{-1} in water, and ~5,000 m s^{-1} in steel.

source-focused illumination

A mode of illumination in a light microscope where the light source, e.g. bulb filament or arc, is focused at the specimen plane. Source-focused or critical illumination is rarely used in light microscopy as Köhler illumination is now preferred.

Sparrow criterion

A criterion for resolving power of an optical system. The Sparrow criterion states that two point objects are resolved when there is the minimum intensity difference between their overlapping Airy disks.

spatial coherence

The degree of coherence of waves across the width of a beam. In a spatially coherent beam one point on the wavefront can interfere with any other equivalent position. Laser light, star light and electron beams from field-emission guns have high spatial coherence since all waves emanate from a point source. Spatial coherence determines brightness since incoherent waves can destructively interfere reducing beam brightness. In an electron microscope the spatial coherence of the beam is approximately equal to the width of the Fresnel fringes formed at an edge, with large defocus.

spatial difference technique

The acquisition of spectra from two different areas of the specimen or from two different areas of a spectrometer detector under otherwise identical conditions. Subtracting one spectrum from the other permits discrimination of small differences in signals that arise from chemical or bonding differences between the two locations (e.g. between a defect and a region of perfect crystal), or permits the removal of spectral artifacts due to defects e.g. in one channel of a diode array in a PEELS spectrometer.

spatial filter
1. An optical device that changes the spatial distribution of a beam.
2. An aperture mask that modulates the passage of light through an optical system.
3. An image filter that changes the spatial frequencies in an image.

spatial frequency
The number of objects per unit length. The spatial frequency of a wave is the reciprocal of wavelength.

spatial resolution
The smallest distance between two clearly separated points or lines in an image.

specific refraction increment see specific refractive increment

specific refractive increment
The change in refractive index of a solution for a 1% increase in solute concentration.

specimen airlock
The airlock used to introduce a specimen into a vacuum device.

specimen block
The hardened embedding medium containing a specimen for microtomy.

specimen chamber, of SEM
The chamber at the base of the column of a scanning electron microscope that houses the stage and detectors.

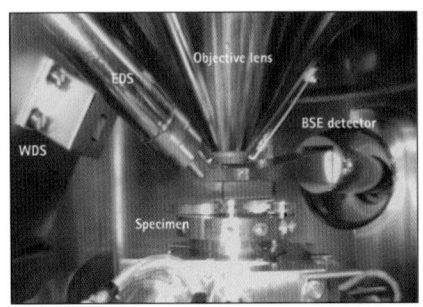

Specimen chamber of SEM showing objective lens, stage and detectors. Courtesy of Carl Zeiss SMT.

specimen drift
The movement of the specimen during observation on a microscope. Common causes of specimen drift include heating by the beam of the specimen or its support, inadequate mechanical or thermal contact between specimen and the holder or stage, and instabilities of the stage.

specimen height
The position of the specimen in the objective lens

of a transmission electron microscope.

specimen holder, for TEM

A rod on which a specimen is mounted for examination in a transmission electron microscope. A typical specimen holder comprises a metal rod with a flattened tip containing receptacles for grids or bulk specimen supports and a device for holding the specimens in place. Specialized holders may have devices for in-situ experimentation.

specimen holder airlock

The airlock used to introduce the specimen holder into the column of a transmission electron microscope.

specimen plane see object plane

specimen processing

The procedures used to prepare a specimen for examination in a microscope.

specimen processor

A device that automates any sequence of steps in the processing of one or more specimens for microscopy.

specimen processor, life sciences see tissue processor

specimen processor, materials science

An apparatus that automates several of the standard procedures used in the processing of materials science specimens for light and electron microscopy. Typical procedures include plasma cleaning, planarization by ion-beam etching, and coating by ion-beam sputttering.

specimen rod see specimen holder
specimen support grid see grid
spectral data cube see spectral image cube

spectral image

An image generated from a 2D slice through a spectral image cube. The intensity and/or color of each pixel indicates the value of the selected

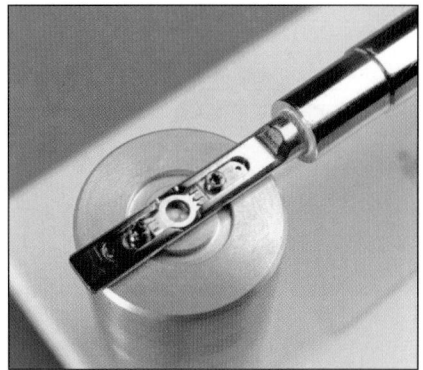

Specimen holder for transmission electron microscope. Courtesy of E.A. Fischione Instuments, Inc.

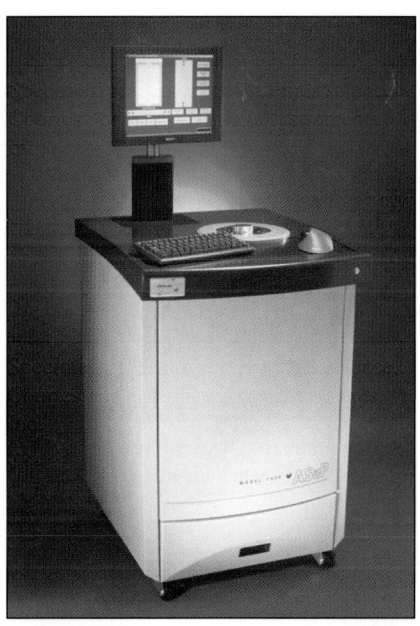

Specimen processor for materials science. Courtesy of E.A. Fischione Instruments, Inc.

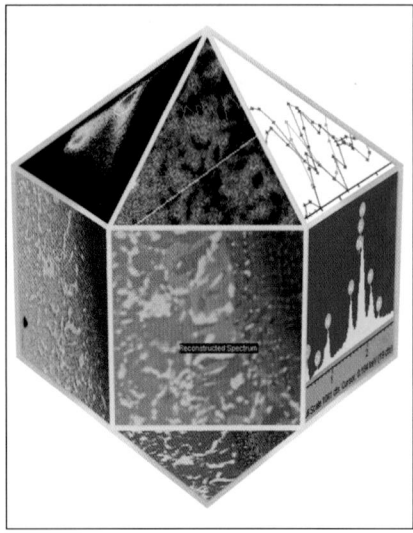

Spectral image cube of WDX data.
Courtesy of Oxford Instruments.

emission at that point in the specimen.

spectral image cube

A three-dimensional graphical representation of the data collected by spectral imaging.

spectral imaging

The collection of a full spectrum of the radiation emitted from each pixel in the specimen using a spectroscopic technique. Spectral imaging creates a 3D spectral data or image cube where the X and Y axes are pixel co-ordinates and the Z axis is wavelength or energy. The spectral data cube can be sampled in any plane to create a spectral image.

spectral linescan

A spectrum obtained by a one-dimensional scan of a specimen by an investigating probe

spectral overlap

The overlap of data within or between excitation, emission or absorption spectra. Spectral overlap may occur in fluorescence microscopy of a multiply labelled specimen when the emission maximum of one fluorophore overlaps the excitation maximum of another.

spectral reflectance

The reflection of a specific wavelength of light.

spectral unmixing

The analysis and separation of the spectrum of wavelengths emitted from fluorescent objects. Spectral unmixing is typically applied to specimens that have been labeled with two or more fluorophores, and requires a spectrometer or spectroscopic analyzer to measure the intensity of each wavelength of light emitted from each pixel in a fluorescence image. In a typical spectral unmixing system, fluorescence emission is dispersed by a glass prism to 32 photomultiplier tubes, each measuring a 10-nm bandwidth of the emission. The data are used to build a spectral image cube from which software, using reference spectral data, can generate a 2D image showing the spatial distribution of fluorescence in each object.

Spectral unmixing by spectroscopy using diffraction efficiency enhancement system. Courtesy of Nikon.

spectrometer

An instrument that contains a mechanism to produce a spectrum of the wavelengths or energies of radiation and charged particles and produce a spectrum.

spectrometry

The measurement of the electromagnetic radiation (and charged particles) with which physical systems interact or that they produce as a means of obtaining information about the systems and their components [IUPAC, 1997].

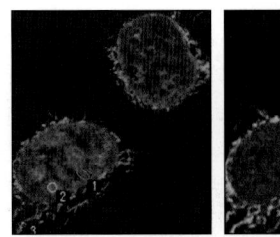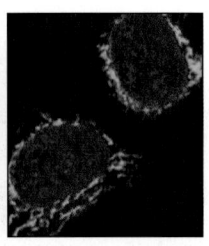

Spectral unmixing. Left: full color image. Right: spectrally unmixed image. Courtesy of Nikon.

spectrophotometer

An instrument that measures absorption spectra. A spectrophotometer has a light source emitting radiation from ultraviolet to infrared, a monochromator to select wavelengths and a photodetector to measure the intensity the light transmitted through the specimen.

spectroscope

An optical instrument that produces a spectrum of the wavelengths of light. A simple spectroscope has a collimator facing the incident light, a prism to disperse the light and an eyepiece to examine the spectrum.

spectroscopic photoemission low-energy electron microscopy

The analysis of photoelectrons released from the surfaces of a specimen in a low-energy electron microscope.

spectroscopic ruler see fluorescence resonance energy transfer

spectroscopy

The study of physical systems by the electromagnetic radiation with which they interact or that they produce [IUPAC 1997]. A term commonly used synonymously with spectrometry.

spectrum

The range of wavelengths, frequencies or energies of radiation, or the range of properties of an entity.

spectrum image see spectral image

specular reflection

The reflection of light with no change in the spatial distribution of the constituent rays.

speed of light

Symbol: c. The speed of light and electromagnetic radiation in a vacuum is 299,792,458 m s^{-1}.

sphere of reflection see Ewald sphere

spherical aberration

An aberration of lenses arising from the dependence of focal length on aperture for non-paraxial rays. In spherical aberration rays from object points on the optical axis passing through the central and peripheral regions of the lens are focused at different focal planes. Spherical aberration is the single most common cause of image degradation in electron microscopy.

spin polarized low-energy electron microscopy

The analysis of the polarization of photoelectrons released from the surfaces of a specimen in a low-energy electron microscope.

spinning disk confocal microscope

A confocal microscope in which the specimen is illuminated through pinhole or slit apertures in a spinning disk.

split photodiode

A photodiode with two light sensing regions allowing the output signals to be compared. Split photodiodes are used to detect the motion of optical cantilevers in scanning probe microscopes.

s polarized

Radiation polarized normal to a plane of incidence.

spot scanning

The consecutive positioning of a probe at preselected positions (spots) on a specimen. Spot scanning is used in confocal microscopy, low-dose electron imaging, cryoelectron microscopy and X-ray microanalysis.

spot size

The diameter of a focused beam in a microscope. In a confocal microscope the spot size is set by the pinhole aperture. In an electron microscope the spot size is set by condenser lens 1.

spray freezing

The cryofixation of particulate specimens by spraying an aqueous suspension into a cryogen.

spring clips

1. Stage clips.
2. Circlips.
3. Devices for retaining grids in the specimen holder of a transmission electron microscope.

spring constant

A measure of the flexibility of the cantilever of a scanning probe microscope. Symbol k; units N m^{-1}. For a beam cantilever:

$$k = Ewt^3/4l^3$$

where E is Young's modulus, w is width, t is thickness, and l is length. Spring constants of typical commercial cantilevers are in the range 0.01 to 200 N m^{-1}.

Spurr's resin

A low viscosity epoxy resin suitable for infiltration into dense specimens such as minerals, bone and plant cells.

sputter coater

An apparatus for the metal coating of specimens by the plasma sputtering technique. A sputter coater consists of a vacuum chamber containing a cathode (the target), typically made of a heavy metal such as Au, Cr, or Pd, an anode and a (cooled) platform for the specimen. An inert gas, usually Ar, is introduced into the evacuated chamber, ionized by a high voltage applied between cathode and anode, and its ions attracted to the cathode where they cause sputtering of target atoms. Target atoms are randomly distributed throughout the chamber by collisions with ionized gas, and so condense on the specimen from all directions ensuring a uniform coating.

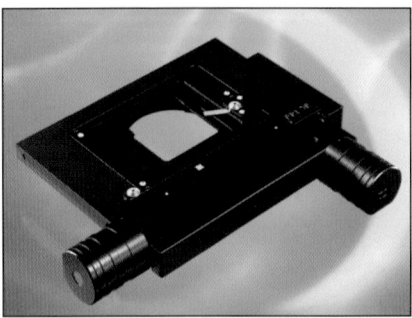

Motor-driven stage for light microscope.
Courtesy of Prior Scientific Instruments.

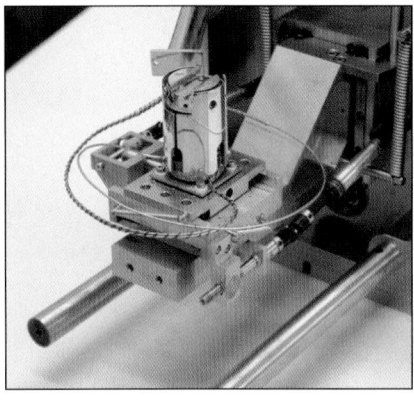

Helium temperature stage for scanning
electron microscope. Courtesy of Gatan,
Inc.

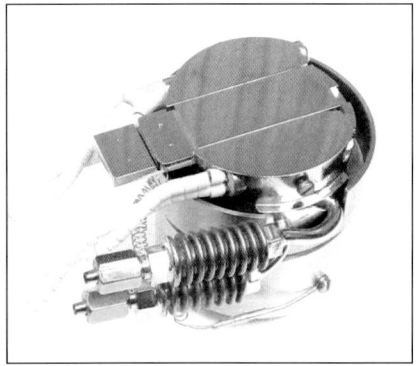

Variable temperature stage for scanning
electron microscope. Courtesy of Gatan,
Inc.

sputter coating

The coating of a specimen with material sputtered
from a target. Sputter coating is commonly used to
prepare specimens for examination in a scanning
electron microscope.

sputtering

The release of atoms from the surface of a
material, typically by bombardment of a cathode
with ionized particles from a plasma or ion beam.

stage

The platform on which the specimen or its support
is placed and held for viewing by a microscope.

stage, of light microscope

The standard stage on a light microscope has a
spring-loaded arm to hold the specimen or slide
and a rack and pinion mechanism to allow
translation of the specimen in x and y directions.
Typical enhancements include mechanisms for
centering, rotation, levelling and motor drive.

stage, of SEM, helium temperature

A stage for the examination of specimens at liquid
helium temperatures (~4K). Cooling to 4K can
enhance cathodoluminescence and change
electronic properties of materials that can be
revealed by other SEM modes.

stage, of SEM, variable temperature

A stage with heating and cooling capabilities for
imaging and analysis of specimens in a scanning
electron microscope at a range of temperatures.
Cooling is achieved with liquid helium or nitrogen,
or with a Peltier plate; heating with resistive
elements. Variable temperature control of a
specimen within the SEM may be used to protect
or enhance its properties or for investigation of
temperature dependent effects. Cooling with liquid
nitrogen (77K) is typically used to protect
specimens from electron beam damage; Peltier
cooling (~250K) is used to stabilise wet specimens
in environmental SEMs. Heating (up to 1500K) is
often used for dynamic investigations of phase
changes or mechanical properties.

stage clip

A flat spring that holds a slide in place on a light microscope stage.

stage drift

The movement of the stage of a microscope during the acquisition of an image or spectrum. Stage drift is a potential problem in all microscopes and is typically caused by heat transfer from incident radiation or mechanical imperfections of stage components.

stage micrometer

A slide engraved with a scale with divisions in millimeters and micrometers. A stage micrometer is typically used in conjunction with an eyepiece reticle to determine image magnification and measure objects.

stage scanning

1. In a scanning probe or confocal light microscope, moving the stage (specimen) relative to a fixed probe or beam in order to form a raster image.
2. In a light or electron microscope, moving the stage in a series of steps to collect a set of images.

Stage micrometers. Courtesy of Agar Scientific.

stain

A molecule or dye that is attached to a specimen to enhance its contrast, add color, or to identify a specific component.

stain precipitates

An artifact of staining procedures commonly seen in transmission electron microscopy of thin sections. Stain precipitates appear as electron dense particles and crystals on the surface of the section. Possible causes include precipitates in the stain, old stain solution, and reactions of stain with air (carbon dioxide) or with specific tissue components.

staining device

A device for batch processing of histological slides or electron microscope grids through a staining process.

stand see microscope stand

standing-wave microscopy

A mode of fluorescence microscopy that uses standing waves to excite fluorescence. In standing-wave microscopy the specimen is placed in a standing wavefield created by the intersection of two coherent, equal amplitude laser beams; fluorescence excitation is maximal from those regions of the specimen that lie at the antinodes of the standing wavefield.

state of polarization

The form of polarization of light, e.g. linear, circular, elliptical.

static mode

The use of an atomic force microscope without oscillation of the cantilever and probe. In static mode, the tip is scanned across the specimen surface either in contact mode or non-contact mode with a constant cantilever deflection due to equilibrium of the attractive long-range force between tip and specimen and the repulsive force at the contact and the bending force from the cantilever. A feedback loop keeps the cantilever deflection constant as the tip moves over the specimen; the feedback signal is used to generate an image of the topography of the specimen.

steel knife

A knife made from steel used to cut thick sections on a microtome.

STEM-TEM see TEM-STEM

steradian

SI derived unit of solid angle. Symbol sr.

stereo

1. Abbreviation for stereoscopic.
2. Relating to three dimensions.

stereology

A method in image analysis that deduces the parameters of a whole specimen from images of a set of sections. In a typical stereological analysis, images of a small number of random sections of the specimen are examined and parameters such as areas, lengths and number of specified objects are quantified using statistical sampling. The data are then used to determine the distribution of the measured parameters in the whole specimen.

stereomicroscope

A binocular light microscope that presents different angular views of the specimen to each eye, simulating normal stereoscopic vision. Modern stereomicroscopes employ either the Greenough or the common main objective designs.

stereomicroscopy

The use of a stereomicroscope

stereo pair

Two images of the same specimen taken from different angles. For viewing with a stereoscope the images are placed side by side or one above the other; alternatively the two images can be merged to form an anaglyph or polarized stereoscopic image.

Stereomicroscope. Courtesy of Carl Zeiss Ltd.

stereopsis see stereoscopic vision

stereoscope

1. A stereo viewer.
2. Abbreviation for a stereromicroscope.

stereoscopic image

A composite image formed by superimposition of the two images of a stereo pair.

stereoscopic microscope see stereomicroscope

stereoscopic vision

The perception by the brain of depth in a scene due to the horizontal displacement of the images formed on each retina of the eyes.

stereoscopy

The methods for viewing the three-dimensional structure of a specimen.

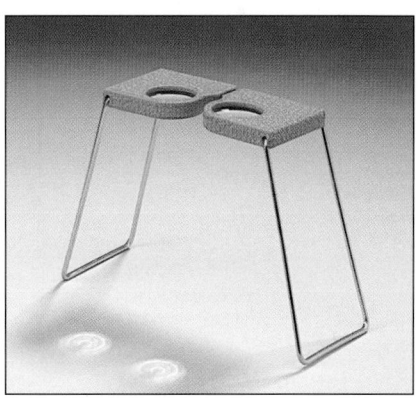

Stereoviewer. Courtesy of Agar Scientific.

stereoviewer

A device for viewing stereo pairs of images. Stereoviewers present each eye with only one of the two images, simulating stereoscopic vision.

stress birefringence see **photoelasticity**
stigmatic see **anastigmatic**

stigmator

A multipole magnetic lens that corrects
astigmatism in an electron microscope. High-
resolution electron microscopes typically have
three stigmators: the condenser stigmator under C2
makes the focused beam symmetrical; the
objective stigmator, under O1, corrects high-
magnification images, and the diffraction
stigmator under D1, corrects diffraction patterns
and low-magnification images.

stimulated-emission depletion microscopy

A technique used in far-field fluorescence
microscopy to improve resolution. In stimulated
emission depletion two ultrafast synchronized
pulsed laser beams are used for excitation: the first
excites fluorescence in the object, whereas the
second, longer wavelength pulse immediately
quenches fluorescence from the outer regions of
the object. STED can improve the point-spread
function threefold horizontally and sixfold axially.

Stokes shift

The reduction in frequency (and increase in
wavelength) of light due to inelastic scattering. The
Stokes shift that occurs in the activation of a
fluorophore can be estimated from the difference
between its excitation and emission maxima.

stop

1. Synonym for a diaphragm, usually of fixed
dimension.
2. An acidic solution that stops the (alkaline)
development of photographic film.

strain-free objective

An objective for polarized light microscopy with
lenses made from glasses free of optical anisotropy
due to strain. A strain-free objective is marked P or
pol.

straining stage

A stage with a mechanism to allow tension or
compression to be applied to the specimen.

streak camera

A camera that detects high-speed changes in image intensity by scanning the image across a detector. In a streak camera the image is focused on a photocathode; emitted electrons are deflected by a fast electrical pulse towards a position sensitive film or CCD detector with femtosecond resolution.

streaking see smear

striae

Linear variations in refractive index caused by imperfections in the manufacture of a glass lens.

stripping film

The layer of film emulsion placed over a specimen in the autoradiography technique.

structure factor

A measure of the atomic scattering factors of the atoms in a unit cell of a crystalline specimen.

structure image

A lattice image formed in electron-optical conditions which are independent of the sample, and which can be directly interpreted as a faithful representation of the projected potential of the sample to some limited resolution. The weak phase-object approximation at Scherzer focus is one condition for obtaining a structure image.

structured illumination

Microscope illumination that is non-uniform or spatially modulated, typically as a result of placing an aperture mask in the illumination path or by using interfering beams. The use of structured illumination in combination with image processsing can improve lateral and axial resolution and remove out-of-focus fluorescence from images.

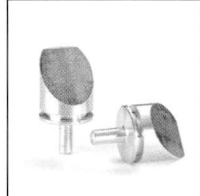

stub

A platform on which specimens are placed for examination in a microprobe or scanning electron microscope. Stubs are usually manufactured from aluminum for most imaging applications, or from low atomic number materials, such as Be or C, for microanalytial applications. The specimen is

Stub holder and stubs for scanning electron microscopy. Courtesy of Agar Scientific.

attached to the stub with an adhesive or conductive paint.

stub holder

A holder for one or more specimen stubs during coating procedures.

student microscope see laboratory microscope

sublimation

The direct change of state from solid to gas. A process used in freeze drying and freeze etching of frozen specimens.

substitution solvent

An organic solvent that replaces the water of a cryofixed specimen during freeze substitution. Commonly used substitution solvents are acetone and methanol.

subtractive colors

Colors produced by the subtraction of one color from white. The subtractive primary colors are cyan, magenta and yellow. Any set of subtractive colors when added together produce black.

superelastic scattering

Scattering with a gain in energy. Superelastic scattering can occur when an electron collides with and de-excites an excited atom.

superresolution

The improvement of the resolution of the images formed by a diffraction-limited light microscope. Methods for superresolution aim to reduce the point-spread function and include such techniques as 4pi microscopy, I5M, structured illumination, and stimulated emission depletion.

SuperTwin lens see Twin lens

support film

A very thin layer of material used to support a specimen for microscopy, typically made of plastic, carbon or silicon.

support grid

A thin mesh or perforated foil used to support specimens for transmission electron microscopy. Support grids are manufactured in many configurations, e.g. mesh, hole, slot, and from a variety of materials, such as copper, gold, nylon. Standard grid diameters are 2.3 and 3.05 and 2.3 mm. A support grid may be coated with a support film.

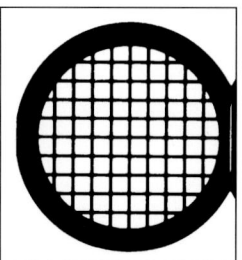

Support grid for electron microscopy. Courtesy of Agar Scientific.

support stand see **microscope stand**
suppressor cap see **Wehnelt**
surface cast see **replica**
surface contour microscopy see **profilometry**
surface electron energy-loss spectroscopy see **reflection electron energy-loss spectroscopy**

surface enhanced Raman microscopy

Raman microscopy in which Raman scattering is enhanced by placing the specimen on a metal substrate.

surface force balance

An apparatus used to investigate the physical properties of fluid samples confined between two closely apposed atomically smooth surfaces.

surface magneto-optical Kerr effect see **Kerr effect**
surface potential microscopy see **scanning surface potential microscopy**
surface replica see **replica**

surround wave

The direct wave that is not diffracted by an object.

S-video

Abbreviation for separate video, a video transmission system that separates the luminance and chrominance signals.

S-wave see **surround wave**
swing-in/out filter tray see **filter tray**
swing-in/out top lens see **top lens**
swing-in/out viewing screen see **focusing screen**

systéme electronique couleur avec mémoire see **SECAM video standard**

T

take-off angle

The angle between the specimen plane and the center of the detector of an X-ray spectrometer. A high take-off angle reduces the chances of absorption of X-rays by the specimen.

Tanaka pattern

A convergent beam electron diffraction pattern containing a single diffraction disk. To form a Tanaka pattern the beam is overfocused and the diffraction aperture is used to select one disk.

tandem scanning confocal microscope

A real-time confocal microscope that uses a Nipkow disk to illuminate and collect light from the specimen. The disk rotates at high speed sending many narrow beams of light to all points of the specimen in each rotation, simultaneously collecting reflected or emitted light through the same apertures (mono scanning) or diametrically opposed apertures (tandem scanning). Enhancements include the use of microlenses to focus light onto the disk.

tapping mode

A mode of operation of an atomic force microscope in which an oscillating probe intermittently contacts the specimen. In tapping mode, the cantilever oscillation amplitude is relatively high; during a small part of the oscillation cycle the tip will experience a repulsive force due to contact with the specimen reducing the oscillation amplitude. A feedback loop keeps the oscillation amplitude constant as the tip scans the specimen; this feedback signal provides the topographic profile of the specimen. The advantages of this mode are the absence of lateral forces between tip and specimen surface, the ability to measure soft materials, and good sensitivity to surface features.

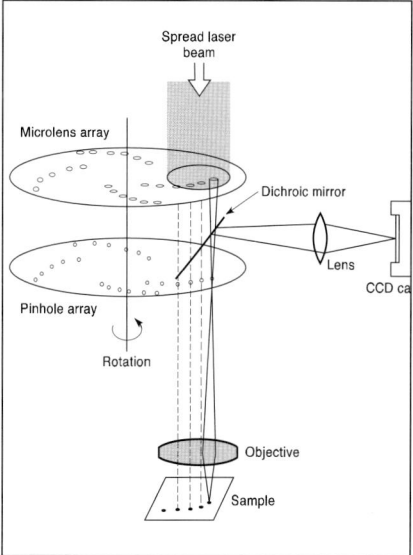

Principle of tandem scanning confocal microscope with Nipkow disk and Yokogawa microlens optics. From D. Murphy ©John Wiley and Sons, Inc.

target
1. The source of coating material in a sputter coater.
2. The light-sensitive structure in a video tube that generates an electrical charge proportional to the intensity at any pixel of the image projected onto the target. The target comprises three layers: a transparent outer electrode at a potential 10-100 V more positive than the gun; a central dielectric; and an insulating photoconductive layer that is scanned by the electron beam.

teaching microscope see multihead microscope
Telan lens see tube lens

telemicroscopy
The operation of a microscope from a remote location via the internet. In telemicroscopy all the microscope operations are fully automated and are controlled by software at the remote location.

telephoto lens
A magnifying lens system containing a positive lens group followed by a negative lens group with an effective focal length longer than the system length.

telepresence microscopy see telemicroscopy

television
1. The transmission and reception of video by radio waves or cable.
2. A monitor capable of receiving television.

TEM-STEM see scanning transmission electron microscope

temporal coherence
The distance over which a beam of waves is coherent. A laser has a temporal coherence length of many meters; light from an incandescent bulb has a very short coherence length (~ 1 µm).

thermal atomic force microscopy see scanning thermal microscopy

thermal diffuse scattering

The inelastic scattering of incident electrons as a result of phonon excitation in the specimen. Thermal diffuse scattering accounts for the diffuse background intensity in an electron diffraction pattern.

thermal drift see specimen drift, stage drift

thermal field emitter see Schottky field emitter

thermal field-emission microscope

A field-emission microscope that uses a thermal field emitter.

thermal field-emission see Schottky field emission

thermal probe

A scanning probe microscope tip or cantilever that detects temperature. Types of thermal probe include thermocouples, resistive elements, bimetallic cantilevers, and conductive probes in which contact potential is temperature dependent.

thermal stage drift see stage drift

thermal-advance microtome

A microtome that uses resistive heating of the specimen arm to advance the specimen towards the knife.

thermionic emission

The emission of electrons produced by heating.

thermionic emitter

A filament that releases electrons when heated.

thermionic gun

A standard type of electron gun used in scanning and transmission electron microscopes. A thermionic gun has three major components: a thermionic source; a Wehnelt; and an anode plate. The filament is held at a negative potential (the accelerating voltage) relative to the anode which is at ground potential.

thermionic source

A filament that is heated to overcome the work function required to release electrons.

thermoelectric cooling see **Peltier effect**
thermoelectric effect see **Peltier effect**

thermoluminescence

Luminescence produced by heat.

thick lens

A lens in which its thickness is significant compared to its focal length, having two principal planes that are not coincident. Most light microscope objectives contain thick lenses.

thickness contrast see **mass-thickness contrast**

thickness map

An image that shows the variation in thickness of a thin specimen. A thickness map can be generated from a electron energy-loss spectrum since the low-loss intensity and the plasmon-loss peak intensity are a function of thickness. In X-ray microanalysis the X-ray characteristic peak intensity is a function of thickness, so X-ray data can also be processed to give thickness maps.

thickness meter see **film thickness meter**

thick section

A section more than 1-μm thick, typically used for light microscopy.

thin-film transistor display

A liquid crystal display in which each pixel has a semiconductor transistor that controls the electrodes and hence the intensity of the pixel.

thin foil see **foil**

thin lens

A lens in which its thickness is small compared to its focal length, having two principal planes that are are coincident.

thin-lens equation

A formula relating object s_o and image s_i distances and focal length f of a thin lens in air by the expression:

$$1/s_o + 1/s_i = 1/f.$$

thin section

A general term for sections cut on an ultra-microtome.

third harmonic generation

A non-linear optical phenomenon in which three identical incident photons combine in a medium to form a single photon with a frequency triple that of the interacting photons.

third harmonic microscopy

The use of third harmonic photons to form an image in a light microscope.

third-order aberrations see monochromatic aberrations

third-order optics

The study of optical systems with consideration of non-paraxial rays and the effects of aberrations. Third-order optics uses as its approximation the first two terms in the expansion:

$$\sin \theta = \theta - \theta^3/3! + \theta^5/5! - \theta^7/7! + \ldots.$$

Thon rings

High intensity concentric rings that occur in a diffraction pattern or Fourier transform of an image. Thon rings are a consequence of the contrast transfer function of the instrument and their parameters are a measure of instrument performance: defocus (ring placement), astigmatism (elliptical rings), drift (incomplete rings) and resolution (ring range).

three-photon excitation

The excitation of a fluorophore by simultaneous absorption of three photons having the combined energy of the normal single photon excitation energy.

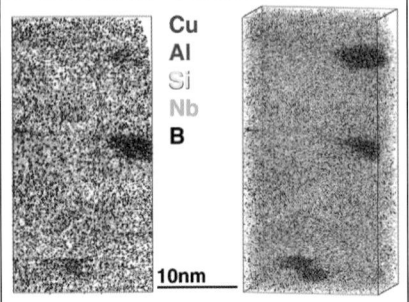

Cu
Al
Si
Nb
B

10nm

3D atom probe and element map of
FINEMET alloy. Courtesy of Oxford
Nanoscience. *Microscopy and Analysis*
2005.

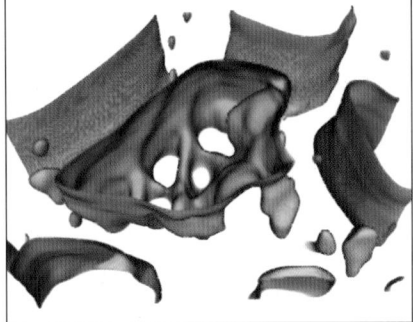

3D reconstruction of a clematis stem cell
and neighboring cell walls, produced from
a Z stack of monochrome autofluorescence
images. After 3D blind deconvolution of
the image stack, the voxel visualization was
created using the blending method.
Courtesy of Soft Imaging System.

three-dimensional atom probe

A type of field-ion microscope that simultaneously
images and identifies atoms in a specimen. The
configuration of a 3DAP is similar to that of a
field-ion microscope except that atoms evaporated
from the tip of the specimen are analyzed one by
one by time-of-flight mass spectrometry with a
position-sensitive detector (PSD). A 3D chemical
image of the tip is reconstructed from the flight
times and impact positions of desorbed atoms at
the PSD. The 3DAP provides the highest spatial
resolution in compositional analysis.

three-dimensional reconstruction

The reconstruction of the structure of a specimen
from a dataset of images acquired: 1. from serial
sections; 2. by tilting the specimen; 3. by imaging
a large number of specimens at different
orientations.

through-focus series

A set of images taken at different planes in a thick
specimen, or at set increments of under- and overfocus of a thin specimen. In a transmission
electron microscope a through-focus series may be
used to obtain one image with optimal focus, or
for phase reconstruction.

TIFF image file format

The tagged image file format widely used for
high-resolution color raster images in microscopy
and publishing. File extensions: .tif, .tiff. The
TIFF format supports RGB, CMYK, grayscale and
uses lossless compression.

tilt rotate holder

A specimen holder that allows both tilt and
rotation of the specimen.

time constant

The time that a pulse processor is allowed to
process an acquired signal. In an X-ray
spectrometer, the time constant determines count
rate and energy resolution: a short constant allows
more counts to be processed; a long constant gives
better resolution but fewer counts.

time-domain fluorescence lifetime

A mode in fluorescence lifetime imaging microscopy in which the fluorophore is excited by a picosecond pulse of laser light and the fluorescence emission is measured at several time points after excitation using a gated detector.

time lapse video

A technique used with video imaging devices to record an event that occurs slowly over time. In time-lapse video the recording frame rate is reduced, e.g. to 1 fps. When the recording is played at normal frame rate, events that are hard to discern by eye become more readily apparent.

tip

The sharp probe that interacts with the specimen in a scanning probe microscope. The tip radius and aspect ratio are important factors in image resolution. SPM tips can be made from a wide variety of materials with many different shapes according to the specific application. For STM, tips are usually made of tungsten, electro-chemically etched to produce an atomically sharp tip. For AFM, tips are usually formed during cantilever microfabrication and may be then sharpened by etching, electron beam deposition or focused ion-beam milling, or by the addition of nanotubes. Surface coatings can be added to tips for additional hardness or conductivity, and functionality can be added by chemical modif-ication or attachment of molecules.

tip artifact see tip convolution
tip broadening see tip convolution

tip convolution

The convolution of the shape of a scanning probe microscope tip with that of a feature on the specimen. Tip convolution is a potential problem in the interpretation of images, particularly when the surface feature is sharper than the probe tip.

tissue-culture microscope

An inverted light microscope with a long working distance condenser and optics designed for observation of cell and tissue cultures in glass and plastic vessels.

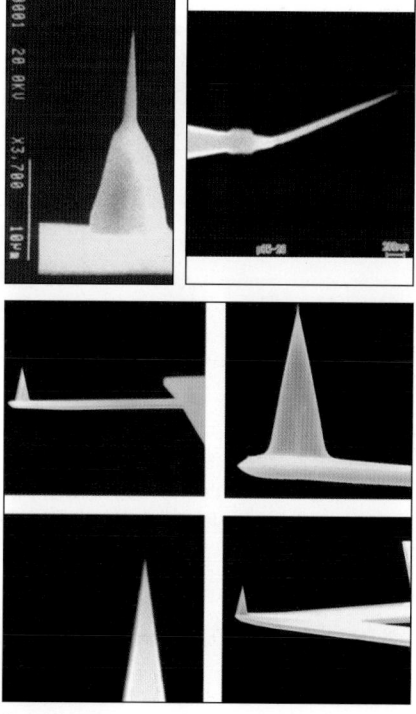

Scanning probe microscope tips. Courtesy of NT-MDT.

Tissue processor for biological samples for transmission electron microscopy. Courtesy of Boeckeler Instruments, Inc.

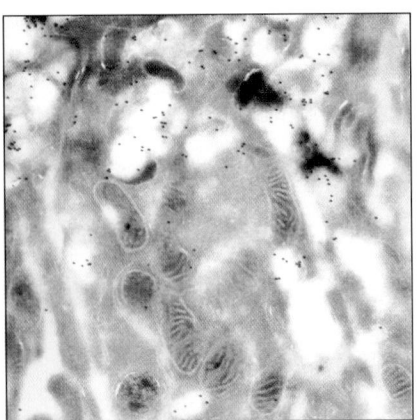

Immunogold labeling of lactase in crysection of pig intestine prepared by Tokuyasu technique. Courtesy of Julian Heath.

tissue processor

A device that automates several of the procedures in the processing of biological specimens for histology and electron microscopy. Tissue processors are typically used after primary fixation of tissues; the processor then takes the specimens through further procedures such as post-fixation, dehydration and embedding.

Tokuyasu technique

A technique for the preparation and staining of ultrathin sections of frozen specimens, typically for immunoelectron microscopy. In the Tokuyasu technique, a lightly chemically fixed specimen is embedded in a cryprotectant such as sucrose, plunge frozen and then thin sectioned on a cryo-ultramicrotome. The thawed sections are positively or negatively stained with heavy-metal salts and air dried. Various modifications of the original Tokuyasu technique are commonly used in TEM immunocytochemistry.

tomogram

An image created using tomography.

tomography

The production of an image of a plane through a specimen. In tomography the three dimensional structure of an object is reconstructed from a set of projection images of different orientations of the object collected either by rotating or tilting a single object or by imaging a large number of objects with differing orientations. An image of any plane through the object can be generated from the 3D reconstruction.

tomography holder see **high-tilt holder**

top-entry holder

A specimen holder that is inserted through the upper bore of the objective lens of a transmission electron microscope.

top-hat filter

An image or spectral filter that separates signals from background. A top hat filter is used to remove the background in an X-ray spectrum. By convoluting the slowly changing bremsstrahlung intensity with a digital function in the shape of a top hat, the Gaussian characteristic peak shapes are left intact but the bremsstrahlung intensity is reduced to zero.

top lens

The uppermost lens of a condenser, nearest the specimen. The top lens is often removable, or temporarily displaceable by a swing-out mechanism, to reduce the numerical aperture of the condenser and allow the use of low magnification objectives.

topography

The physical features of a surface.

topography mode see constant height mode

topomicroscopy

The analysis of topography using two-dimensional image data acquired by a light or scanning electron microscope.

torr

A non-SI unit of pressure. One torr equals 133.32 pascals.

torsional resonance mode

The use of a torsionally resonating cantilever in a scanning probe microscope. In torsional resonance mode the probe is in contact with the specimen; lateral interactions of the tip and specimen change the torsional resonant frequency of the tip.

total internal reflection

The reflection of all incident rays when light travelling through a medium of high refractive index meets an interface with a medium of lower refractive index. Total internal reflection occurs when the angle of incidence exceeds the critical angle.

total internal reflection fluorescence

The excitation of fluorescent molecules by evanescent waves.

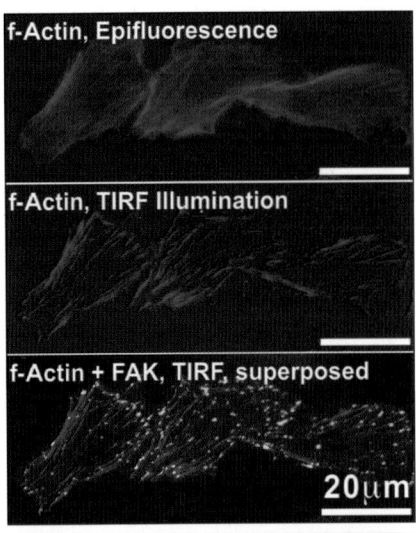

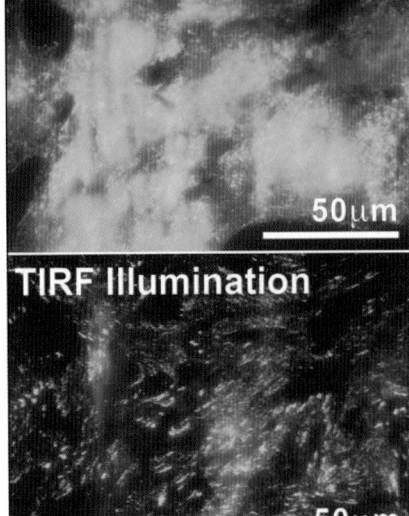

Comparison of standard epifluorescence and total internal reflection fluorescence imaging of fluorescently labeled cells. Courtesy of Götz Pilarczyk, Uta Joos, Till Biskup, Oliver Ernst, Ines Westphal, Hauke Kahl and Olympus.

total internal reflection fluorescence microscopy

A mode in fluorescence microscopy that uses total internal reflection of laser light to generate evanescent wave excitation of fluors at a glass-liquid interface, typically a coverslip carrying living cells. There are two common microscope configurations for TIRF: prism-based systems in which the specimen is illuminated from beneath the objective, and objective-based systems that use epi-illumination. In prism-based TIRF the coverslip is attached with immersion oil to a glass prism which is used to direct laser excitation light towards the specimen at the critical angle. The microscope objective is immersed in the liquid to detect fluorescence. In objective-based TIRF the coverslip is sealed to an oil-immersion objective in the normal way and laser light is directed down one side of the objective; the lens directs excitation light towards the specimen at the critical angle and the objective collects the emitted fluorescence.

total internal reflection fluorescence objective

A high magnification, high numerical aperture (100x, 1.65 NA) apochromatic objective used for objective-based total internal reflection fluorescence microscopy.

total internal reflection fluorescence spectroscopy

The analysis of spectra obtained by total internal reflection fluorescence microscopy.

total transmittance see transmittance

transfer loop

A loop used to transfer specimens, section or grids. The loop is typically formed from platinum wire, plastic or other materials.

transfer noise

Noise produced in a CCD by the transfer of charges from pixels to the output node.

transitional solvent

An organic solvent used in specimen processing to assist the infiltration of media into a specimen.

transmission electron microscope

An electron microscope that uses the electrons transmitted through a specimen for image formation and spectroscopy. The key components of a TEM are: a thermionic or field-emission electron gun; one or more condenser lenses; a goniometric stage; an objective lens; intermediate and projector lenses; a projection chamber with a phosphor screen; and accessory detectors for imaging and spectroscopy.

transmission electron microscopy

The use of a transmission electron microscope.

transmittance

The ratio of the intensity of radiation leaving a specimen to the intensity of the incident radiation. Internal transmittance relates to radiation that enters and leaves the specimen; total transmittance takes into account radiation that is reflected before entering the specimen.

transmitted fluorescence microscopy

The use of transmitted light in fluorescence microscopy. Transmitted fluorescence microscopy is not a widely used technique and requires the use of a high-NA brightfield or darkfield condenser instead of the objective to focus excitation light onto the specimen, and an excitation filter above the objective.

transmitted light

Light that passes through an object.

transmitted-light interference microscope see interference microscope

transverse dynamic force microscopy

The use of a scanning probe microscope to map and measure the transverse forces between the probe and a specimen. In transverse dynamic force microscopy the probe oscillates parallel to the specimen surface. Typical applications for TDFM are studies of biological macromolecules and synthetic polymers.

transverse magnification see lateral magnification

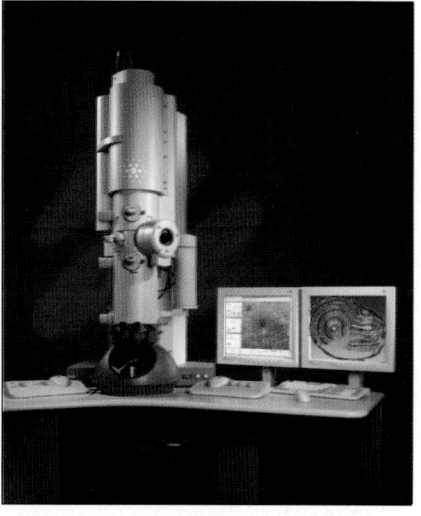

Transmission electron microscope.
Courtesy of FEI Company.

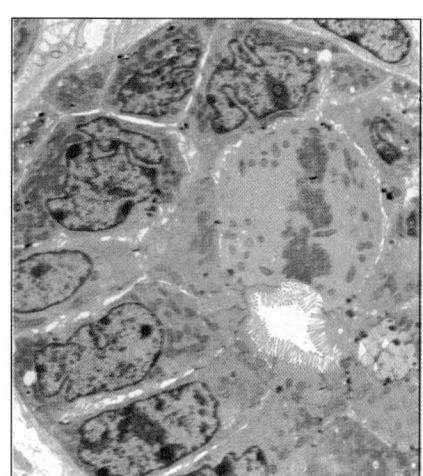

Transmission electron micrograph of ultra-thin section through intestinal crypt cells.
Courtesy of Julian Heath.

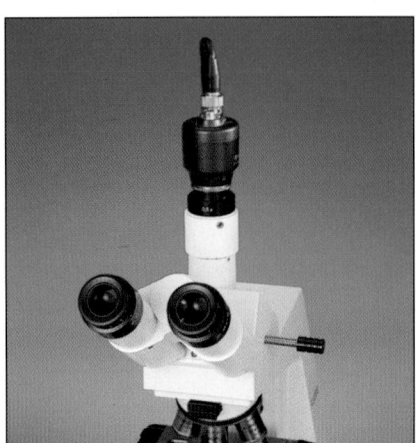

Trinocular viewing head on light microscope. Courtesy of Carl Zeiss Ltd.

Tripod polisher on rotary polisher. Courtesy of Agar Scientific.

triboluminescence

Luminescence produced by friction.

trimming, of block

A procedure for shaping a block for microtomy. Trimming can be done manually, with a razor blade, on a block trimming device, or on the microtome.

trinocular tube see **trinocular viewing head**

trinocular viewing head

A viewing head on a light microscope that carries two eyepieces for the eyes and a projection eyepiece for a detector.

triode gun

An electron gun with three elements: an anode, a cathode and a biased third electrode.

tripod polisher

A device used to cross-section or thin materials by abrasive thinning and polishing. The tripod polisher has micrometer-adjustable feet that allow angled grinding and polishing of specimens on a rotary grinder or polisher. A tripod polisher can produce a wedge-shaped specimen with an electron transparent tip suitable for transmission electron microscopy.

tripod scanner

A scanner with three separate piezoelectric elements that move the specimen or probe in a scanning probe microscope.

trough see **water trough**

tube lens

A lens in the body tube or viewing tube of a light microscope with infinity-corrected optics that collects the parallel light emerging from the objective and forms an intermediate image at the eyepieces.

tube scanner

A cylinder of piezoelectric material, such as PZT (lead zirconium titanate), that moves the specimen or probe in a scanning probe microscope. Three dimensional movement is regulated by voltages applied to the tube. A tube scanner has electrodes attached to each outside quadrant and to the inner surface. When voltages +U and -U are applied to opposite outer electrodes the tube bends producing translation; voltages applied to the inner electrode cause axial extension or shortening.

tube see microscope tube

tubelength

The mechanical tubelength is a prescribed distance from the shoulder of an objective to the top of the eyepiece tube of a light microscope. Older microscopes have a tubelength of around 160 mm. Modern microscopes with infinity corrected optics are not restricted by a tubelength.

tunable filter

An optical filter in which the transmission spectrum can be varied electronically, such as an acousto-optical filter or liquid-crystal filter.

tungsten filament

1. A filament formed from a thin V-shaped tungsten wire, used in the thermionic gun of electron microscopes. The filament is spot welded to two tungsten support wires which are embedded in a ceramic insulator. The melting point of tungsten is 3422°C.
2. A coil of tungsten wire used in incandescent lamps.

Tungsten filaments for electron guns. Courtesy of Agar Scientific.

tungsten single-crystal filament

A field emitter used in cold and Schottky field-emission guns of electron microscopes. A tungsten single-crystal filament is made from a needle-shaped piece of crystalline tungsten welded to the tip of a tungsten filament. The crystal has a flat circular facet normal to the optical axis; for Schottky emission the tip is coated with zirconium oxide which reduces the work function from 4.5 to 2.8 eV.

tuning-fork sensor

A piezoelectric cantilever in the shape of a tuning fork. For use a probe is attached to the face or side of one of the tines and the fork is orientated parallel or perpendicular to the specimen. The tuning fork is operated at its resonant frequency electrically or by attachment to an additional piezo device. Interaction of the probe and specimen changes the amplitude and phase of the tuning fork.

tunneling atomic force microscopy

The use of an atomic force microscope with a dc biased conductive probe to measure and map the local electrical conductivity or electrical integrity of resistive specimens such as semiconductors.

tunneling current

The flux of electrons that occurs between two closely apposed (~1 nm) conductors when a bias voltage is applied. Tunneling current varies exponentially with separation distance.

turbidometry

The measurement of particle density in a suspension by analysis of the scattering of transmitted light.

turbomolecular pump

A vacuum pump that uses a turbine to evacuate a space. A turbomolecular pump contains a set of rotor blades spinning at high speed (up to 50,000 rpm) that strike gas molecules forcing them towards the exhaust region of the pump from where they are removed by a rotary backing pump. Turbomolecular pumps can produce vacuums in excess of 10^{-9} Pa.

turret see condenser turret, objective turret

tweezers

A tool for the handling of specimens, grids and small instruments parts. Tweezers are manu-factured from different materials, such as carbon steel or stainless steel, have different properties, e.g. antimagnetic, non-magnetic, inert, and are available in many configurations, e.g. pointed, cross-over, anti-capillary, to suit their many applications in specimen handling.

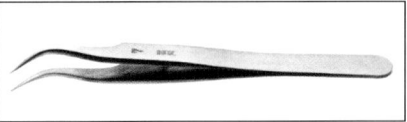

Tweezers for grid and specimen handling. Courtesy of Agar Scientific.

Twin lens

A proprietary design of magnetic objective lens, usually symmetric, containing a minilens in the front bore to provide the flexibility needed for STEM and TEM operation in the same instrument.

two-beam condition

A condition in electron diffraction, where the specimen is orientated to give a diffraction pattern containing the direct beam and one diffracted beam from one set of crystal planes.

two-photon excitation

The excitation of a fluorophore by simultaneous absorption of two photons having the combined energy of the normal single photon excitation energy. For operational reasons, the two excitation photons usually have the same wavelength: e.g. two infrared photons of 700-nm wavelength can stimulate emission from a fluorophore with an excitation maximum of 350 nm.

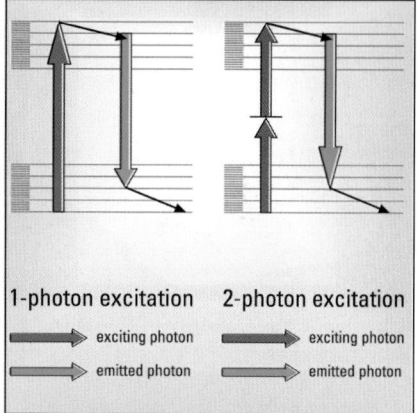

Principle of two-photon excitation. Jablonski diagram showing single and two photon excitation. Courtesy of Leica Microsystems.

U

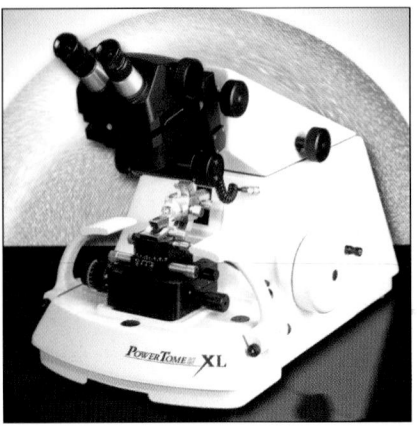

Ultramicrotome. Courtesy of Boeckeler Instruments, Inc.

UHF connector

A standard screw-on connector used to couple co-axial cables to video and other electronic equipment using ultrahigh frequency signals.

ultrahigh vacuum

A gas pressure lower than 10^{-8} pascals.

ultrahigh vacuum AFM or STM

The operation of an atomic force microscope or scanning tunneling microscope under ultrahigh vacuum conditions in order to minimize contamination and permit atomic resolution imaging.

ultramicrotome

A type of microtome for the production of thin and ultrathin sections. Modern ultramicrotomes are fully automated, often under computer control. A motor-driven specimen arm moves the block past the knife edge during the cutting cycle; the knife is retracted by an electromagnet during the return stroke. The specimen advance mechanism is usually mechanical, but some older machines use a thermal advance. Continuously variable section thicknesses (30 nm to 2 μm) and cutting speeds (0.1 to 100 mm s^{-1}) are usually standard. A stereomicroscope and illumination system are used for observation of the specimen and collection of sections.

ultramicrotomy

The cutting of thin sections of a specimen on an ultramicrotome.

ultrasonic disk-cutting device

A cutting device equipped with an ultrasonically vibrating cutting tool. The cutting tool is typically a hollow cylinder of titanium, axially vibrated at ~25 kHz, that cuts a 3-mm diameter disk with the aid of an abrasive slurry.

Ultrasonic disk-cutting device. Courtesy of E.A. Fischione Instruments, Inc.

ultrasonic force microscopy

The use of an atomic force microscope to measure and map the mechanical properties of an ultrasonically vibrated specimen. In ultrasonic force microscopy the specimen is vibrated at sub-nanometer amplitude by placing it on an ultrasound generator, such as a piezoelectric crystal; an AFM tip in contact with the specimen surface can then measure local mechanical compliance.

ultrasound wave

A sound wave that is inaudible to the human ear with a frequency greater than 20,000 Hz.

ultrastructure

The structure of a specimen as revealed by nanometer-resolution images.

ultrathin section

A section 30-100 nm thick cut on an ultra-microtome and suitable for transmission electron microscopy.

ultrathin window

A type of window used in X-ray spectrometers, usually made of a 100-nm thick film of polymer, diamond or silicon nitride.

ultraviolet

Electromagnetic radiation with wavelengths in the range 4-350 nm.

ultraviolet A

Ultraviolet radiation with wavelengths in the range 320-400 nm, generally regarded as harmless to the skin.

ultraviolet B

Ultraviolet radiation with wavelengths in the range 290-320 nm, causing reddening and tanning of the skin.

ultraviolet C

Ultraviolet radiation with wavelengths in the range 230-290 nm, generally regarded as harmful to the skin and a possible cause of skin cancer.

ultraviolet filter

A filter that blocks ultraviolet light and transmits visible light.

ultraviolet microscopy

The use of ultraviolet light as the source of illumination in a microscope.

underfocus

The condition where an image is formed in a plane behind (downstream of) the true or Gaussian image plane.

uniaxial crystal

A crystal having one optic axis and two indices of refraction. Uniaxial crystals are members of the crystallographic groups hexagonal, tetragonal and trigonal; examples are calcite and crystal quartz.

unidirectional shadowing

The shadowing of a stationary specimen at a fixed angle.

unipotential lens see einzel lens

universal condenser

A light microscope condenser designed for use with a wide range of objectives and contrasting techniques, typically having a condenser turret.

unmasking see antigen retrieval
unpolarized light see randomly polarized light

unscattered electrons

Electrons that are not scattered by a specimen and exit with no change in direction or energy. Unscattered electrons form the zero-order spot in an electron diffraction pattern.

unsharp mask

An image filter that sharpens the boundaries between objects of different intensity. In unsharp masking a blurred version of the original image is partially subtracted from the image creating a difference image or mask which is then iteratively added to the original image.

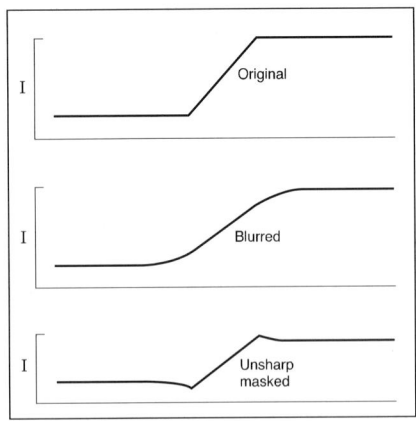

Principle of unsharp masking. From D. Murphy © John Wiley and Sons, Inc.

upright light microscope

The conventional configuration of a light microscope with the illumination source and condenser below the specimen and the objectives and eyepieces above.

uranyl acetate

A fixative and stain widely used in the processing of biological specimens for transmission electron microscopy. Uranyl acetate may be used as a fixative, as a positive stain for tissues and thin plastic sections, or as a negative stain for particles.

V

vacancy
A missing atom in a crystal lattice.

vacuum
A space with, or the state of, low gas pressure. A perfect vacuum would have no molecules but this is impossible to achieve in practice as the materials surrounding any space have a finite vapor pressure.

vacuum bellows
A flexible, accordion-like metal seal used in vacuum systems, e.g. to connect an X-ray spectrometer to the column of an electron microscope.

vacuum desiccator see film desiccator
vacuum evaporator see coating unit

vacuum grease
A hydrocarbon, silicone or synthetic grease used as a sealant and lubricant in vacuum systems.

vacuum seal
An air-tight seal in a vacuum system, typically formed using O-rings, metal seals, vacuum bellows, and/or vacuum grease.

vane pump
A vacuum pump that uses vanes to seal the pumping chamber.

vapor-jet pump see diffusion pump
Varel contrast, Varel optics see variable relief contrast

variable-angle total internal reflection fluorescence microscopy
A mode of total internal reflection microscopy in which fluorescence emission is varied by changing the angle of incident light and hence the depth of the evanescent wave.

variable numerical-aperture objective

A light microscope objective with an internal iris diaphragm that alters the numerical aperture of the lens.

variable-pressure scanning electron microscope

A scanning electron microscope in which the specimen chamber is separated from the column by a pressure-limiting aperture, allowing the examination of specimens in their native state at pressures in the range 2-133 Pa.

variable-pressure secondary electron detector

A secondary electron detector designed for use in a VPSEM. Emitted secondary electrons collide with gas molecules in the specimen chamber creating further electrons by avalanche multiplication; during these collisions photons are emitted which are detected by the VPSE detector.

variable relief contrast

A proprietary design of modulation contrast light microscopy that combines phase contrast and variable unilateral or oblique illumination.

variable self-biased gun see self-biasing gun

V coating

An anti-reflection coating that reduces reflection of one wavelength of light.

vector graphic

A digital image that is stored as a set of geometrical shapes, e.g. points, lines, and polygons, rather than pixels, allowing for smaller file sizes and simple rescaling on any display.

vertical illuminator see epi-illuminator

vibrating knife

A diamond knife equipped with an ultrasonic device that vibrates the knife during sectioning.

vibrating microtome

A type of microtome that has a vibrating steel

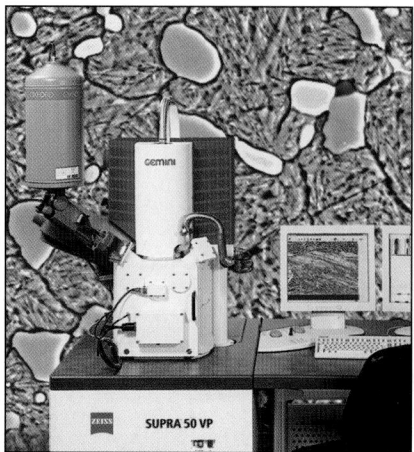

Variable-pressure scanning electron microscope. Courtesy of Carl Zeiss SMT.

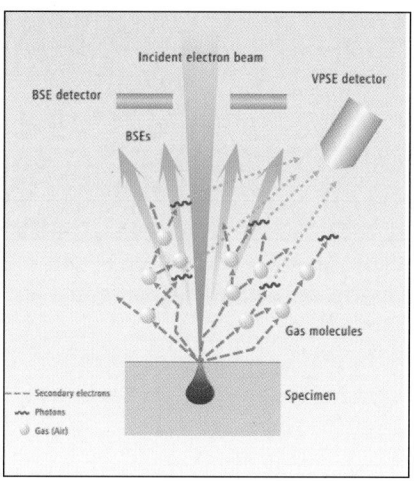

Principle of secondary electron detection in a variable-pressure scanning electron microscope. Courtesy of Carl Zeiss SMT.

knife, typically used for cutting thick sections of delicate, living or lightly fixed specimens.

vibration planes see planes of vibration

video

The electronic recording, broadcasting and displaying of moving images.

video camera

A camera that outputs a video signal. Video cameras contain video tubes or charge-coupled devices and output an analog video signal that can be displayed on a television monitor.

video capture card see frame grabber

video enhancement

The enhancement of image contrast by electronic enhancement and digital video processing. Video cameras can detect intensity differences much smaller than those detectable by the human eye. By adjusting the gain and offset of a video camera, and using digital video processing procedures such as frame averaging and background subtraction, video microscopy can produce high-contrast images in real time from unstained specimens such as living cells.

video frame see frame of video

video microscopy

The use of video cameras to form images and video in light microscopy.

video-rate camera

A video camera outputting frames at the NTSC or PAL standard video rates.

video standards see PAL, NTSC or SECAM video standards

video tube

The imaging device that generates the signal in a video camera. A video tube is an evacuated glass tube with a target that is scanned raster fashion by a beam from an electron gun. When the target is illuminated, the resistance of its photoconductive

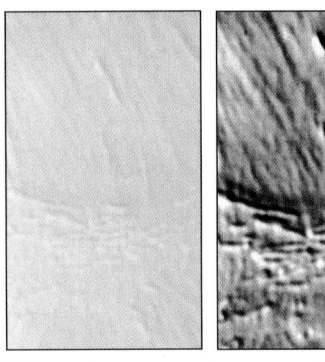

Video enhancement of contrast of DIC image of cell organelles. Courtesy of Carl Zeiss Ltd.

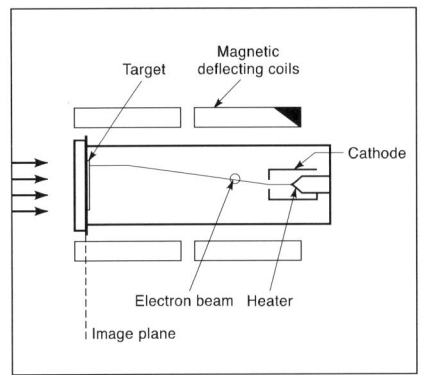

Schematic of vidicon camera tube. From D. Murphy © John Wiley and Sons, Inc.

layer is reduced, allowing a current to flow from target to gun; this current is used to form the video signal.

vidicon tube
A standard type of video tube used in video and television systems.

viewing chamber see projection chamber

viewing head
That part of the tube of a light microscope that carries the eyepieces and detectors; for infinity-corrected optics it may also carry the tube lens.

viewing screen
A phosphor screen used to view an image. In a transmission electron microscope the whole screen, or a central portion, can be inclined at 45° for more comfortable viewing and focusing or raised to 90° to allow electrons to pass to a detector or film plate. The screen has markings that delineate the centre and margins of the detector.

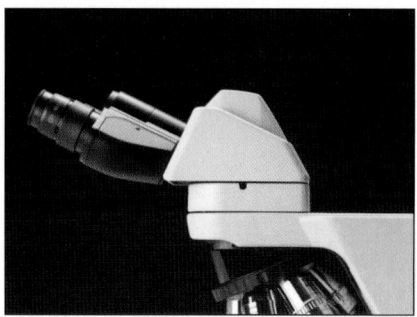
Viewing head of light microscope.
Courtesy of Nikon.

viewing tube see viewing head

viewing window
The lead-glass windows in the viewing chamber of a transmission electron microscope.

vignetting
The loss of detail at the edges of an image caused by an optical component framing or windowing the field of view.

virtual aperture
A notional aperture created in object space by placing an aperture in image space. A virtual aperture is created when using the selected-area diffraction aperture in electron diffraction.

virtual image
An image that cannot be seen on a screen or recorded by a detector. In a light microscope the real image received on the retina appears to be a virtual image formed at a point about 25 cm from the eye.

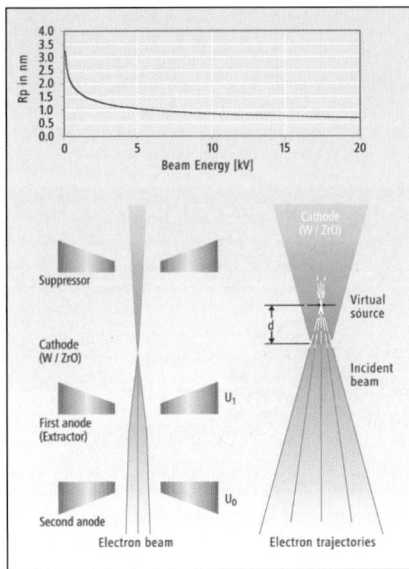

Electron optics of field-emission gun showing virtual source of electrons in a field emittter. Courtesy of Carl Zeiss SMT.

virtual microscopy

The use of a computer to simulate the operation of a microscope, typically for training purposes.

virtual source, of electrons

The point in an electron gun from which electrons appear to originate. For a field-emission gun this point is located inside the filament; for a thermionic gun, it lies near the gun cross-over point.

visible light

Electromagnetic radiation in the visible spectrum.

visible spectrum

The wavelengths of electromagnetic radiation that can be detected by the human eye, typically light of wavelengths 400 nm (violet) to 700 nm (red).

visual astigmatism

Astigmatism of the eye, caused by asymmetries in the curvature of the cornea or the refractive power of the lens.

visual axis

The axis of the eye that passes through the center of the lens and the macula.

visual field number see field number

vitreous humor

The gelatinous fluid filling the posterior chamber of the eye. Refractive index = 1.337.

vitreous ice

Ice with no crystalline structure. In cryoelectron microscopy the presence of vitreous ice in the specimen can be confirmed by examining a diffraction pattern.

vitrification

The formation of vitreous ice. Vitrification of hydrated specimens requires rapid cooling to a temperature less than -138°C at a rate exceeding 10,000°C s^{-1}. Vitrification avoids the damage to specimen structure caused by ice crystal formation.

voltage center see **voltage centering**

voltage centering

An alignment procedure in a transmission electron microscope that ensures that electrons remain on the optic axis as their energy changes. In voltage centering the high-voltage supply is wobbled about the required setting, and the gun tilt controls are used to center the beam.

volume element see **voxel**

voxel

An abbreviation for volume element, the smallest unit of a three-dimensional digital image.

W

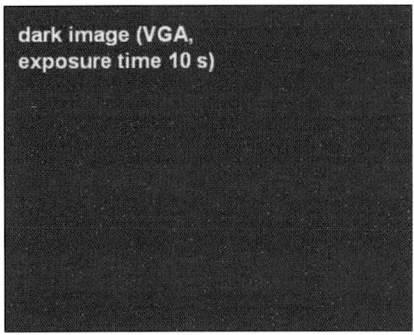

Warm pixels in low-light CCD image.
Courtesy of PCO.

Water chiller for electron microscope.
Courtesy of Agar Scientific.

warm pixel

A high intensity pixel in a CCD image produced by charge leakage. Warm pixels are an inherent artifact of CCDs and are commonly seen in long exposure images of low intensity specimens.

water chiller

A device that supplies cooling water to the lenses of an electron microscope, to vacuum pumps, to coating units, and other instruments that require cooling.

water cooling see cooling systems

water-immersion objective

A light microscope immersion objective designed for use with an aqueous medium, e.g. to observe living cells or hydrated specimens in open dishes.

water trough

A water reservoir for the collection of sections cut on an ultramicrotome. Plastic, metal or tape troughs may be attached to glass knives; diamond knives are available with integral troughs.

water window

The range of X-ray wavelengths or energies least absorbed by water. The X-ray water window lies between the K edges of oxygen (543 eV, 2.34 nm) and carbon (284 eV, 4.4 nm). Within this window more X-rays are absorbed by carbon than water allowing good contrast and penetration of hydrated organic biological and materials specimens in X-ray microscopy.

watt

SI derived unit of power and radiant flux. Symbol W. One watt equals one joule per second.

wave

A periodic disturbance in a medium. Traveling waves may be longitudinal as in sound or transverse as in light, and are characterized by their amplitude, energy, frequency, speed and wavelength.

wavefront

A line or surface joining all points of equal phase in a set of waves.

wavefront reconstruction see holography

wavelength

The distance between successive crests of a traveling wave. Symbol λ. Alternatively, the length of one complete cycle of a wave, or the distance between points of equal phase on adjacent waves.

wavelength-dispersive X-ray fluorescence

X-ray fluorescence using a wavelength dispersive spectrometer.

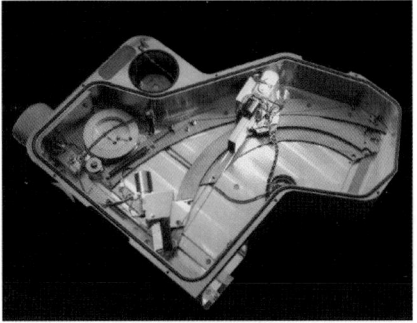

wavelength-dispersive spectrometer

A spectrometer that analyzes radiation by wavelength, typically by using a diffraction grating.

wavelength-dispersive spectroscopy

The use of a wavelength-dispersive X-ray spectrometer.

wavelength-dispersive X-ray spectrometer

A spectrometer that characterizes X-rays by wavelength. The key components of a wavelength dispersive spectrometer are one or more diffracting crystals with known interplanar spacings, an X-ray detector (usually a gas proportional counter), processing electronics, and a computer to control data acquisition, processing and display. To produce a partial X-ray spectrum a crystal is tilted through an angle of ~50°, by moving it around the Rowland circle, allowing X-rays of wavelengths that meet the Bragg condition for that crystal to be diffracted towards the detector. To obtain a full spectrum the process is repeated with several crystals that have different interplanar spacings.

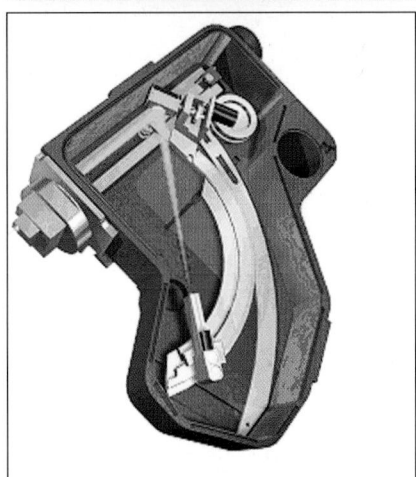

Wavelength-dispersive X-ray spectrometer. Courtesy of Oxford Instruments.

wavelength-dispersive X-ray spectrometry see wavelength-dispersive X-ray spectroscopy

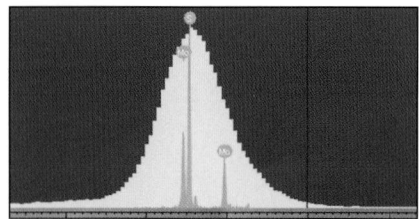

Comparison of energy-dispersive (yellow) and wavelength-dispersive (blue) X-ray spectra of molybdenum sulfide, MoS_2. Only WDX resolves the peaks for Mo and S. Courtesy of Oxford Instruments.

wavelength-dispersive X-ray spectroscopy

The identification of elements by analysis of the wavelengths of the X-rays emitted by excitation of specimen atoms with a focused electron beam. Wavelength-dispersive X-ray spectroscopy of bulk specimens is typically performed in a scanning electron microscope or in a microprobe. Emitted X-rays have a wavelength that is characteristic of the excited element. WDX has a higher energy resolution (peak widths of 2-20 eV) than EDX, giving better sensitivity and is better at resolving lighter elements, but data acquisition is slower. The spatial resolution is a function of the electron probe size at the specimen.

wavelength-dispersive X-ray spectrum

A plot of intensity of emitted X-rays versus wavelength (or energy) generated by wavelength-dispersive X-ray spectroscopy.

wavelength-ratioing microscopy see fluorescence ratioing

wave number

SI derived unit for wave frequency. Symbol k. Wave number is the reciprocal of wavelength.

wave optics

A method in optics that describes the propagation of light through optical systems in the form of waves with consideration of the phenomena of coherence, diffraction and interference.

wave plate see retardation plate

wavetrain

A sequence of identical waves, having equal phase, amplitude, wavelength and polarization.

weak-beam darkfield microscopy

A mode of darkfield electron microscopy, typically used to study dislocations in materials, in which the the objective aperture is placed around one weakly diffracted beam that is used to form the image.

Wehnelt

A cylindrical metal cap with an aperture that surrounds the filament of thermionic and field-emission guns. The Wehnelt carries a bias voltage that regulates the intensity of electron emission and the area of the filament from which it occurs.

Wehnelt aperture see **Wehnelt**
Wehnelt bias see **gun bias**
Wehnelt cap see **Wehnelt**
Wehnelt cylinder see **Wehnelt**
Wehnelt voltage see **gun bias**

wet scanning electron microscopy

The examination of hydrated specimens in a scanning electron microscope.

wetting agent

A chemical that reduces surface tension, aiding the adsorption and spreading of a specimen on a substrate.

white light

Light having a mixture of wavelengths (colors) that enables it to be perceived as white by the eye.

white radiation see **bremsstrahlung**

white-light interference microscopy

A form of interferometric microscopy that uses polychromatic light as the illumination, typically used for surface profilometry.

widefield eyepiece

A light microscope with a high field-of-view number.

Wien filter

An electron energy filter containing crossed magnetic and electrostatic fields.

winding(s)

The copper wire wound inside a magnetic lens.

windowless detector

An X-ray spectrometer that operates without a window, allowing all X-rays emitted by the specimen to reach the detector.

window support grid

A thin grid of silicon or tungsten wires that supports the window of an X-ray spectrometer.

wobbler

A mechanism on an electron microscope that serves as an aid in alignment and focusing. A wobbler operates by cyclically varying the currents in a lens, a set of beam deflectors or the accelerating voltage, creating a 'wobbling' or oscillating image of the beam or specimen on the screen.

Wollaston prism

A beamsplitting prism used in differential interference contrast microscopes. A Wollaston prism contains two wedges of crystal quartz or calcite with orthogonal optic axes cemented together at the hypotenuse. A ray of polarized light passing through the prism emerges as divergent E and O rays with orthogonal states of polarization.

work function

The potential difference or energy required to extract an electron from a material. Symbol: Φ; unit: eV. The work function can be overcome by heating to high temperatures (thermionic filaments) or by application of an external electric field (field-emission filaments). For thermionic (W) and cold-field emitters, $\Phi = 4.5$ eV; for LaB_6 filaments, $\Phi = 2.7$ eV; for thermal field-emission filaments, $\Phi = 2.8$ eV.

working distance

1. The distance between a probe and a specimen.
2. In light microscopy, the distance between the front surface of the objective and the coverslip (or specimen if unmounted) when the specimen is in focus.

wrinkles

An artifact of sections caused by creasing during cutting or drying. Section wrinkles can be minimized by spreading floating sections with heat or solvents.

X

XBO lamp see xenon arc lamp

xenon arc lamp
A type of arc lamp filled with high pressure xenon. A xenon arc lamp emits a continuous spectrum of UV, visible and IR light with intense peaks in the IR region.

X-ray
Electromagnetic radiation with wavelengths in the range 10^{-9} to 10^{-11} m and energies in the range 10 eV-100 keV. X-rays are classified as either hard or soft X-rays according to energy, wavelength and properties.

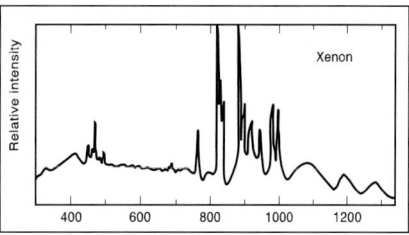

Emission spectrum of xenon arc lamp. From D. Murphy © John Wiley and Sons, Inc.

X-ray collimator
An aperture in front of an X-ray spectrometer that restricts entry of background X-rays and backscattered electrons. The collimator is typically made of tantalum coated with carbon to reduce secondary X-ray emission.

X-ray count rate
The number of X-rays detected by a detector per unit time. In X-ray microanalysis the X-ray count rate is proportional to X-rays emitted and hence to probe current.

X-ray crystallography
The study of the three dimensional arrangement of atoms in crystalline materials by X-ray diffraction. X-ray crystallography is commonly used to investigate biological macromolecules, such as proteins and nucleic acids, which are crystallized and then placed in an X-ray diffractometer where the diffraction pattern is recorded on film or a X-ray detector.

X-ray diffraction
The diffraction of X-rays by the atoms in a crystalline specimen.

341

X-ray energy-dispersive spectrometer/ spectroscopy see energy dispersive X-ray spectrometer/spectroscopy

X-ray fluorescence

The emission of X-rays as a result of excitation by X-rays. In X-ray fluorescence an incident X-ray ejects an electron from an atomic inner shell creating a vacancy that is filled by an electron from a higher energy shell with the release of a secondary X-ray photon.

X-ray fluorescence microscope

A microscope that uses X-ray fluorescence to generate images and spectra of a specimen.

X-ray fluorescence spectroscopy

The elemental compositional analysis of a specimen using X-ray fluorescence.

X-ray mapping

The generation of an image showing the spatial distribution of an element in a specimen using X-ray microanalysis. Each pixel in the X-ray map has an intensity or color proportional to the concentration of the selected element at that position in the specimen. X-ray maps are acquired by scanning the electron probe in raster fashion over the specimen, dwelling at a selected time at each location to collect X-ray signals.

X-ray microanalysis see microanalysis

X-ray microscope

A microscope that uses X-rays to form an image. A typical X-ray microscope uses X radiation from a synchrotron source or from a X-ray tube; a condenser or light guide focuses the radiation onto the sample. A zone plate acts as an objective lens and produces an enlarged image which is detected by a CCD camera.

X-ray microscopy

The use of X-rays to image and investigate a specimen.

X-ray microscopy, in SEM see X-ray ultramicroscope

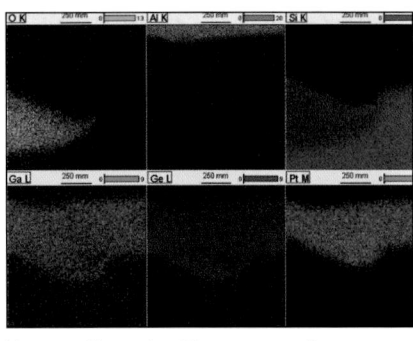

Energy-dispersive X-ray maps of cross-sectioned semiconductor. Courtesy of Thermo Electron Corporation.

X-ray microtomography

Microtomography using X-rays as the source of illumination.

X-ray photoelectron spectroscopy

A technique for compositional analysis using soft monochromatic X-rays to eject surface photoelectrons.

X-ray photoemission electron microscopy

The use of photoelectrons, emitted from the specimen by excitation with X-ray sources, for imaging and spectroscopy.

X-ray shielding see radiation shields

X-ray spectrometer see energy-dispersive spectrometer, wavelength-dispersive spectrometer

X-ray spectrum

A plot of intensity of emitted X-rays versus energy or wavelength.

X-ray take-off angle see take-off angle

X-ray tomography

Tomography using X-rays as the source of illumination.

X-ray tube

A source of X-rays for X-ray diffraction and microscopy. An X-ray tube typically comprises a heated tungsten filament (the cathode) and a metal target (the anode) in an evacuated glass tube. Electrons from the filament are accelerated by an applied high voltage towards the target, causing emission of Kα X-rays which exit through a window in the tube. The wavelength of emitted X-rays is determined by the target which is typically made of chromium (emitting 0.23 nm X-rays), cobalt (0.18 nm), copper (0.15 nm), iron (0.19 nm) or molybdenum (0.07 nm).

X-ray ultramicroscope

A proprietary design of X-ray microscope built inside a scanning electron microscope. X-rays generated by a target positioned under the electron

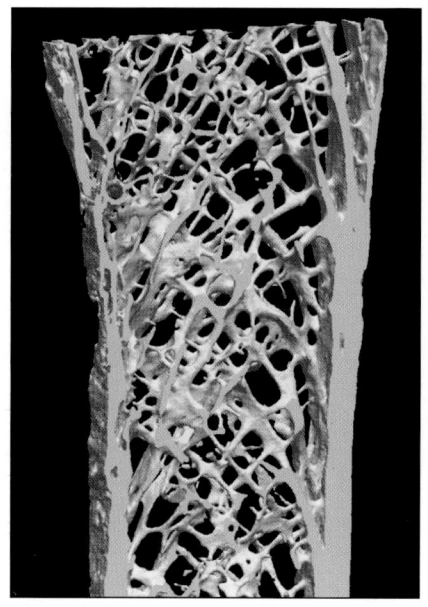

X-ray microtomography of human iliac bone. *Microscopy and Analysis* 2005.

beam are used to form a projection image of the specimen on a CCD camera.

Y

Y-C video see S-video

Y-C video see S-video

yellow fluorescent protein see fluorescent proteins

Young's fringes

The interference fringes produced by Young's slits or a homologous system. Young's fringes are used to estimate the information limit of transmission electron microscopes and are produced by shifting the specimen during exposure; in this case the slits are the spaces between periodic objects.

Young's slits apparatus

A device used to demonstrate the diffraction and interference of radiation. Light is passed through two sets of apertures: the first containing a single slit and the second two adjacent slits; the emergent waves produce an interference (diffraction) pattern of parallel light and dark fringes whose spacing is proportional to the separation of the second pair of slits and the wavelength of light used.

yttrium aluminum garnet

A crystalline compound used as a scintillator ($e_{max} \cong 560$ nm) in electron microscope detectors.

yttrium aluminum perovskite

A crystalline compound used as a scintillator ($e_{max} \cong 378$ nm) in light and electron microscope detectors.

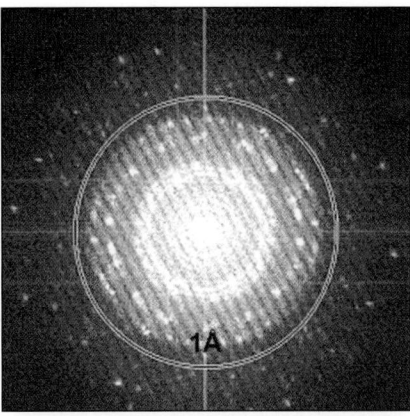

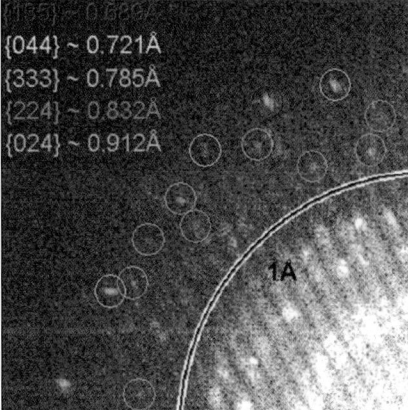

{044} ~ 0.721Å
{333} ~ 0.785Å
{224} ~ 0.832Å
{024} ~ 0.912Å

Young's fringes experiment of poly-crystalline gold, showing the information limit of a high resolution transmission electron microscope. Courtesy of FEI Company.

Z

Z

1. Symbol for atomic number.
2. Symbol for height.

Z contrast

Image contrast that derives from the atomic number of atoms in the specimen. Z contrast is the primary mechanism of image formation in atomic-resolution high-angle annular darkfield scanning transmission electron microscopy, and in backscattered scanning electron microscopy.

Z correction

1. Atomic number correction (see ZAF correction).
2. The adjustment of the height of a specimen.

ZAF correction

A calibration procedure used in quantitative X-ray microanalysis of bulk specimens that accounts for atomic number Z, absorption A, and X-ray fluorescence F. The ratio of the concentrations of an element in a specimen C_i to that in a pure sample of that element $C_{[i]}$ is given by:

$$C_i/C_{[i]} = ZAF.I_i/I_{[i]}$$

where I_i and $I_{[i]}$ are the X-ray intensities emerging from the specimen and standard respectively.

zero-loss electrons

Electrons that have minimal energy loss after scattering by a specimen.

zero-loss filtered imaging

The use of zero-loss electrons to form an image in an energy filtering transmission electron microscope. By removing inelastically scattered (energy loss) electrons, chromatic aberration is minimized, improving spatial resolution. Zero-loss filtered imaging is useful when imaging thick specimens.

zero-loss peak

A peak in an electron energy-loss spectrum formed by electrons with zero energy loss. The width of the zero loss peak is determined by the inherent energy spread of the electron probe.

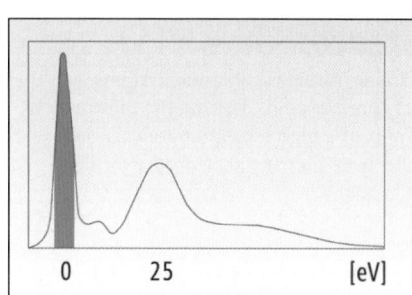

Zero-loss peak of EELS spectrum.
Courtesy of Carl Zeiss SMT.

zero-order Laue zone see **Laue zones**

zero-order light

Light that is not deviated by a specimen.

zero-order maximum

The maximum intensity of the undiffracted photons or electrons in the central spot of a diffraction pattern.

zone axis

The direction common to two or more crystal planes. If an electron beam is parallel to a zone axis, the spots in the diffraction pattern are reciprocally related to the crystal plane spacing.

zone of confusion see circle of least confusion

zone plate

A device used for focusing a beam of X-rays, consisting of a set of concentric apertures of radially decreasing width (down to ~30 nm) in a metal plate that diffract the beam to a focus. The zone plate is used as an objective lens in X-ray microscopy.

zoom

A continuous or stepped change in magnification.

zoom lens

An adjustable lens system that increases magnification without changing the positions of the object and image planes. Zoom lens magnifiers are often placed in the body tube of light microscopes and stereomicroscopes and typically in two configurations: discontinuous zoom, with a rotating drum containing separate sets of fixed magnification lenses; and continuous zoom, consisting of one set of three lenses whose relative positions can be changed variably.

zoom optic

An optical system that zooms an image, typically placed between a camera and either the viewing head of a light microscope or an objective lens.

zoom stereomicroscope

A stereomicroscope equipped with zoom optics.

Zoom optics. Courtesy of Prior Scientific Instruments.

BIBLIOGRAPHY

This bibliography lists the primary sources I have used to compile and write the *Dictionary of Microscopy*. These sources include books, journals, websites, and brochures and technical information from microscopy businesses. I have used many other sources, in print and on the web, but unfortunately they are too numerous to mention, so I would like to take this opportunity to thank the originators of these sources and acknowledge their valuable contribution to this dictionary.

Acoustical Imaging Vol. 26, edited by R. Maev. Kluwer Academic/Plenum Publishers, NY, USA, 2002.

Applied Scanning Probe Methods, edited by B. Bhushan, H. Fuchs, S. Hosaka. Springer-Verlag: Berlin, 2004.

A Practical Guide to Scanning Probe Microscopy, by R. Howland and L. Benatar. Park Scientific Instruments, 1996.

Artifacts in Biological Electron Microscopy, edited by R. Crang and K. Klomparens. Plenum Press, New York, USA, 1988.

Basics of Holography, by P. Hariharan. University of Cambridge Press, UK, 2002.

Biomedical Electron Microscopy, by A. Maunsbach and B. Afzelius. Academic Press, 1999.

Biophotonics International. Laurin Publishing Co., Pittsfield, MA, USA.

Characterization Techniques of Glasses and Ceramics, edited by J. Rincon and M. Romero. Springer-Verlag, Berlin, 1999.

Characterization of Materials, Vols 1 and 2, edited by E. Kaufmann. Wiley-Interscience, NJ, USA, 2003.

Confocal and Two-Photon Microscopy: Foundations, Applications and Advances, edited by A. Diaspro. Wiley-Liss, New York, USA, 2002.

Confocal Laser Scanning Microscopy, by Carl Zeiss, 2003.

Dictionary of Light Microscopy, Royal Microscopical Society Nomenclature Committee, edited by S. Bradbury et al. Royal Microscopical Society Microscopy Handbook Series, 15, Oxford University Press, 1989.

Electron Diffraction in the Transmission Electron Microscope, by P. Champness. Royal Microscopical Society Handbook Series, 47, Bios Scientific Publishers, Oxford, UK, 2001.

Electron Energy Loss Spectroscopy, by R. Brydson. Royal Microscopical Society Microscopy Handbook Series, 48, Bios Scientific Publishers, Oxford, UK, 2001.

Electron Microscopy, by J. Bozzola and L. Russell. Jones and Bartlett, Boston, USA, 1992.

Energy-Dispersive X-Ray Analysis in the Electron Microscope, by A.J. Garratt-Reed and D.C. Bell. Royal Microscopical Society Handbook Series 49, Bios Scientific Publishers, Oxford, UK, 2003.

Field Guide to Geometrical Optics, by J. Greivenkamp. SPIE Field Guide, Vol FG01, SPIE, Bellingham, USA, 2003.

Fundamentals of Crystallography, 2nd Edition. Edited by C. Giacovazzo. IUCr Texts on Crystallography 7, Oxford Science Publications, UK, 2002

Fundamentals of Light Microscopy and Electronic Imaging, by D.B. Murphy. Wiley-Liss, New York, 2001.

Gold and Silver Staining: Techniques in Molecular Morphology, edited by G. Hacker and J. Gu. CRC Press, New York, USA, 2002.

Handbook of Biological Confocal Microscopy, edited by J. Pawley. IMR Press, Madison, WI, USA, 1989.

Hands-on Morphological Image Processing, by E. Dougherty and R. Lotufo. SPIE Press, Bellingham, WA, USA, 2003.

High-Resolution Electron Microscopy, 3rd Edn, by J. Spence. Monographs on the Physics and Chemistry of Materials 60, Oxford University Press, UK, 2003.

High-Resolution Imaging and Spectrometry of Materials, edited by F. Ernst & M. Ruehle. Springer-Verlag, Berlin, Germany, 2003.

HyperPhysics Website: http://hyperphysics.phy-astr.gsu.edu/hbase/hph.html

Imaging in Neuroscience and Development: A Laboratory Manual, edited by R. Yuste and A. Konnerth. Cold Spring Harbor Laboratory Press, MA, USA, 2005.

Introduction to Conventional Transmission Electron Microscopy, by M. De Graef. Cambridge University Press, UK, 2003.

Introduction to Scanning Transmission Electron Microscopy, by R.J. Keyse, A.J. Garratt-Reed, P.J. Goodhew, and G.W Lorimer. Royal Microscopical Society Handbook Series, 39, Bios Scientific Publishers, Oxford, UK, 1998.

Journal of Electron Microscopy, Oxford Journals, UK.

Journal of Microscopy, Blackwell Publishing Ltd, Oxford, UK.

**Large-Angle Convergent-Beam Electron Diffraction (LACBED): Applications to Crystal

Defects, by J-P. Morniroli. Monograph of the French Society of Microscopies, Paris, France, 2002.

Light Microscopy in Biology: A Practical Approach, 2nd Edn, edited by A. J. Lacey. Oxford University Press, 1999.

Live Cell Imaging: A Laboratory Manual, edited by R. Goldman and D. Spector. Cold Spring Harbor Laboratory Press, MA, USA, 2005.

Low Temperature Methods in Biological Electron Microscopy, by A. Robards and U. Sleytr. Series editor: Audrey Glauert, Elsevier, The Netherlands, 1985.

Matter, University of Liverpool Website: http://www.matter.org.uk/matter.htm

Methods in Cellular Imaging, edited by A. Perisamy. American Physiology Society, Oxford University Press, UK, 2001.

Microscopic Techniques in Biotechnology, by M. Hoppert. Wiley-VCH, Weinheim, Germany, 2003.

Microscopy and Analysis. John Wiley and Sons, Ltd, UK

Microscopy and Microanalysis. Cambridge University Press, UK

Microscopy from the very beginning, by Carl Zeiss.

Microscopy, Immunohistochemistry, and Antigen Retrieval Methods for Light and Electron Microscopy, by M. Hyatt. Kluwer Academic/Plenum Publishers, New York, USA, 2002.

Molecular Expressions: Exploring the World of Optics and Microscopy Website: http://micro.magnet.fsu.edu

Nikon MicroscopyU: Explore the World of Optics and Imaging Technology Website: http://www.microscopyu.com

Olympus MIC-D: Digital Microscope Website: http://www.mic-d.com/index.html

Optical Imaging and Microscopy: Techniques and Advanced Systems, edited by P. Török and Fu-Jen Kao, Springer Series in Optical Sciences, Springer-Verlag, Berlin, 2003.

Optics (4th Edn), by E. Hecht. Addison Wesley, San Francisco, USA, 2002.

Optics and Optical Instruments Catalog, by Edmund Optics, 2004.

Oxford Guide to X-ray Microanalysis, CD-ROM Tutorial, by Oxford Instruments, High Wycombe, UK, 1988.

Plant Microtechnique and Microscopy, by S.E. Ruzin. Oxford University Press, 1999.

Practical Application of Light, by Melles Griot, 1999.

Practical Guide to Scanning Probe Microscopy, by Park Scientific Instruments (Veeco), 1996.

Practical Stereology, 2nd Edn, by J. Russ and R. Dehoff. Kluwer Academic/Plenum Publishers, New York, 2000.

Proceedings of Microscopy and Microanalysis Conferences, 1998-2005. Microscopy Society of America.

Review of Scientific Instruments. American Institue of Physiscs.

Resin Microscopy and On-Section Immunocytochemistry, 2nd Edn, by G. Newman and J. A Hobot. Springer Lab Manuals, Springer-Verlag, Berlin, 2001.

Scanning Probe Microscopes: Applications in Science and Technology, by K.S. Birdi. CRC Press, Boca Raton, FL, 2003.

Scanning Probe Microscopy and Spectroscopy: Theory, Techniques and Applications, 2nd Edn, edited by D. Bonnell. Wiley-VCH, New York, 2001.

Scanning Probe Microscopy: The Lab on a Tip, by E. Meyer, H. Hug, R. Bennewitz. Springer-Verlag, Berlin, 2004.

Techniques in Immunocytochemistry, Vols 1-3, edited by G. Bullock and P. Petrusz. Academic Press, 1982-1985.

Transmission Electron Energy Loss Spectrometry in Materials Science and the EELS Atlas, 2nd Edn, edited by C. Ahn. Wiley-VCH, Weinheim, 2004.

Transmission Electron Microscopy. A Textbook For Materials Science, by D. Williams and C. Carter. Plenum Press, New York, 1996.

Understanding Materials: A Festschrift for Sir Peter Hirsch, edited by C. Humphreys. Maney Publishing, London, 2002.

Vacuum Deposition of Thin Films, by L. Holland. Chapman and Hall, London, 1970

Video Microscopy, by Shinya Inoué, Plenum Press, New York, 1986.

Wikipedia Website: http://en.wikipedia.org

X-ray Microanalysis for Biologists, by A. Warley. Practical Methods in Electron Microscopy Vol. 16, Series editor: Audrey Glauert. Portland Press, London, 1997.

Sponsor Information

BAL-TEC AG

BAL-TEC AG
Neugruet 7, P.O. Box 62
FL 9496 Balzers
Principality of Liechtenstein
www.bal-tec.com

Boeckeler Instruments, Inc.

Boeckeler Instruments, Inc.
4650 S. Butterfield Drive
Tucson AZ 85714-3404
USA
www.boeckeler.com

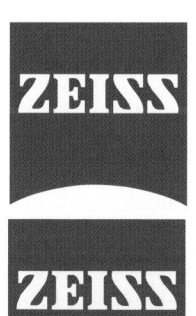

Carl Zeiss Ltd.

Carl Zeiss Ltd.
PO Box 78
Welwyn Garden City
Herts, AL7 1LU
UK
www.zeiss.co.uk

Carl Zeiss SMT - Nano Technology Systems Division

Carl Zeiss SMT
Carl-Zeiss-Str. 56
73446 Oberkochen
Germany
www.smt.zeiss.com/nt

E.A. Fischione Instruments, Inc.

E.A. Fischione Instruments, Inc.
9003 Corporate Circle
Export, PA 15632
USA
www.fischione.com

FEI Company

FEI Company
PO Box 80066
5600 KA Eindhoven
The Netherlands
www.feicompany.com

Gatan, Inc.

Gatan Corporate Headquarters
5933 Coronado Lane
Pleasanton, CA-94588
USA
www.gatan.com

 **Hamamatsu Photonics**

Hamamatsu Photonics
2 Howard Court
10 Tewin Road
Welwyn Garden City
Herts, AL71BW, UK
www.hamamatsu.co.uk

 **Leica Microsystems (UK) Ltd**

Leica Microsystems (UK) Ltd
Davy Avenue,
Knowlhill
Milton Keynes, MK58LB
UK
www.leica-microsystems.com

 **ETS-Lindgren**

Lindgren RF Enclosures, Inc.
400 High Grove Blvd.
Glendale Heights
IL 60139
USA
www.lindgrenrf.com

 **Nikon Instruments**

Nikon Instruments
Schipolweg 321
1171 PL Bad Hoevedorp
1170 AE Bad Hoevedorp
The Netherlands
www.nikon-instruments.com

 **NT-MDT**

NT-MDT
Building 167
Zelenograd
124460 Moscow,
Russia
www.ntmdt.com

 **OLYMPUS LIFE AND MATERIAL SCIENCE EUROPA GMBH**

OLYMPUS LIFE AND MATERIAL
SCIENCE EUROPA GmbH
Wendenstr. 14-18
D-20097 Hamburg
Germany
www.olympus-europa.com

Oxford Instruments

Oxford Instruments
Halifax Road
High Wycombe
Bucks, HP12 3SE,
UK
www.oxford-instruments.com

PCO AG

PCO AG
Donaupark 11
93309 Kelheim
Germany
www.pco.de

Prior Scientific

Prior Scientific Instruments Ltd, 3-4
Fielding Ind. Estate Wilbraham Rd
Fulbourn,
Cambridge, CB15ET, UK
www.prior.com

Renishaw plc

Renishaw plc
Old Town
Wotton-under-Edge
Gloucestershire GL127DW UK
www.renishaw.com

Soft Imaging System

Soft Imaging System
Johann-Krane-Weg 39
D-48149 Münster
Germany
www.soft-imaging.net

StockerYale Inc.

StockerYale Inc.
32 Hampshire Road
Salem, NH 03079
USA
www.stockeryale.com

Thermo Electron Corporation

Thermo Electron Corporation
5225 Verona Road
Madison, WI 53711
USA
www.thermo.com

WITec GmbH

WITec GmbH
Hörvelsinger Weg 6
D-89081 Ulm
Germany
www.WITec.de